HVAC for Beginners

The Ultimate Guide to Heating, Ventilation, and Air Conditioning Systems | Comprehensive Installation, Troubleshooting, and Repair for Residential & Commercial

Connor Wells

Copyright

Copyright © 2024 Connor Wells

All rights reserved.

No portion of this book may be reproduced in any form without written permission from the publisher or author, except as permitted by U.S. copyright law.

Contents

Introduction: Ensuring Comfort and Efficiency   1
1. Introduction to HVAC Systems   9
2. Residential HVAC Installation   19
3. Refrigeration Cycle in HVAC   31
4. Heat Transfer in HVAC   45
5. Troubleshooting HVAC Systems   55
6. HVAC System Controls   68
7. Energy Efficiency in HVAC   80
8. Air Quality in HVAC Systems   90
9. HVAC Maintenance Practices   100
10. Innovations in HVAC Technology   114
11. Heating Systems in HVAC   126
12. Cooling Systems in HVAC   138
13. HVAC System Design Principles   151
14. Emergencies and Safety in HVAC   164
15. Commercial HVAC Systems   177
16. Career Paths in HVAC   189
17. Regulatory Compliance in HVAC   202
18. Case Studies in HVAC   215
19. The Future of HVAC   228
20. Practical Application   242

Golden Rules of HVAC   256

Conclusion	263
Glossary	265
Bonus Chapter: Fun Facts About HVAC	271
Bonus Quiz	275

Introduction: Ensuring Comfort and Efficiency

You wake up on a frosty winter morning, only to find that your home's heating system has failed. The biting cold seeps through every corner of the house, and no number of blankets can fend it off.

This scenario is all too familiar for many homeowners who have experienced the collapse of their Heating, Ventilation, and Air Conditioning (HVAC) systems at the most inopportune moments. While such episodes are inconvenient and uncomfortable, they also raise a critical question about the reliability and maintenance of these essential home systems.

Problems with HVAC systems often stem from neglect, inappropriate orientation or lack of proper maintenance culture. Homeowners might not realize the importance of regular upkeep until something goes wrong. This kind of neglect results in costly repairs or replacements, as evidenced by the millions spent annually on HVAC services across the U.S.

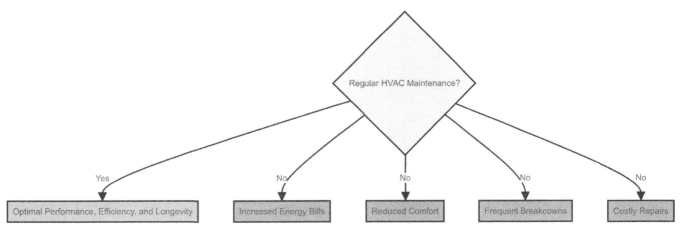

The Domino Effect of Neglecting HVAC Maintenance

Beyond financial implications, poorly maintained systems can also pose safety risks. Malfunctioning heating units can lead to severe health hazards like hypothermia during freezing temperatures. Also poor maintenance practices in central cooling systems can lead to air borne disease and breathing difficulties caused by components like dirty and contaminated air-bags and filters in air handling units etc. Other scenarios, such as faulty air conditioning units causing fires, further underscore the significance of keeping these systems in optimal condition.

Having a reliable and well-maintained HVAC system is vital. So, let's get started on this journey towards mastering HVAC essentials for a better home experience.

The Critical Role of HVAC Systems

Maria had always prided herself on being a diligent homeowner. She made sure her house was in top shape, from the roof to the foundation. Yet, one fateful winter night, as temperatures plummeted well below freezing, the family's biggest nightmare came true: their heating system failed.

With two young children and an elderly parent at home, the situation quickly escalated into an emergency. They huddled together under layers of blankets, watching their breath misting in the frigid air, while they desperately awaited the arrival of an emergency HVAC technician.

Unfortunately, Maria's story is not unique. Each year in the U.S., about three million heating and cooling systems need to be replaced, leading to a staggering $14 billion spent on HVAC services or repairs (U.S. Department of Energy and ENERGY STAR®). This often happens because many homeowners underestimate the importance of regular HVAC maintenance and reliable systems until it's too late.

Over half of a typical home's energy usage stems from heating and cooling (EIA - Energy Information Administration, n.d.). When these systems are poorly maintained or inefficient, the consequences aren't just limited to increased energy costs—they can also compromise safety.

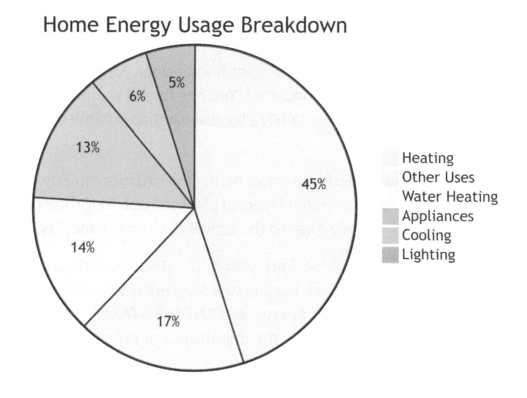

Where Your Home's Energy Goes

For instance, a malfunctioning furnace in cold weather isn't just uncomfortable; it can lead to health issues such as hypothermia, especially for vulnerable populations like children and the elderly.

Not taking care of your HVAC systems can cause more problems than just pain and money problems. If you don't maintain or install your HVAC system properly, it can lead to terrible or even fatal results. This shows how important it is to keep your system in good shape.

The first step toward avoiding such incidents is understanding the value of investing in high-efficiency HVAC equipment. According to the U.S. Department of Energy, replacing outdated equipment with high-efficiency models can cut energy consumption by 50% for electric heating and cooling systems and 10% for gas furnace heating systems (Residential Program Guide). When properly installed, ENERGY STAR® certified heating and cooling equipment can yield annual energy bill savings of 10-30%, without sacrificing functionality.

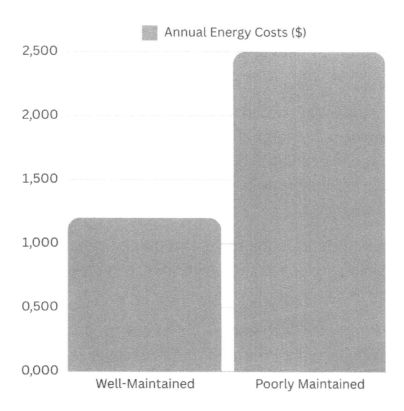

The Financial Impact of HVAC Maintenance

Nevertheless, investing in energy-efficient machinery is just half the battle. Installation practices play a crucial role in the effectiveness of HVAC systems. Improper installation can increase energy use by up to 30% (Residential Program Guide).

If you want your installation done right, hire experts that adhere to tried-and-true methods, including those set down by the Consortium for Energy Efficiency (CEE). Promoting dependable maintenance services such as planned preventive maintenance, having standardized efficiency norms, and using excellent installation procedures are all highlighted in these guidelines.

While trusting professionals is essential, homeowners should also be proactive in maintaining their HVAC systems. Regular maintenance tasks/ routine inspects like changing filters, checking refrigerant levels, and ensuring ductwork is sealed can make a huge difference.

The ENERGY STAR® program suggests that up to 20% of conditioned air gets lost due to leaks, holes, and disconnections in duct systems. Sealed ductwork ensures that the air reaches its intended destinations, enhancing overall efficiency and comfort.

Technologies like air-source heat pumps (ASHPs) have become more common in colder locations. But, if not adjusted for these situations, conventional techniques of sizing and installation might result in inefficiencies. To make sure these systems work well even in very cold weather, the Northeast Energy Efficiency Partnerships (NEEP) have put together comprehensive instructions to help installers. Customer satisfaction and optimal system performance can be achieved with the help of these resources, which offer clear instructions.

Systemic developments in HVAC systems offer further advancements beyond individual households. The Department of Energy is supporting innovative projects that are leading to more efficient and eco-friendly solutions. There are efforts underway to drastically cut energy use, such as the creation of membrane-based rooftop air conditioners and the development of sophisticated heat exchangers.

The investment in high-efficiency equipment and quality installation pays off in multiple ways—reduced energy bills, prolonged system lifespan, enhanced safety, and minimized risk of emergency breakdowns.

Ultimately, the responsibility lies with each of us to ensure our homes are equipped with reliable HVAC systems. By staying informed and proactive, we can create a safer, more comfortable living environment and avoid the chilling scenarios faced by homeowners like Maria.

Common Challenges and Frustrations

Dealing with HVAC systems can be challenging. High energy bills, uncomfortable indoor temperatures, and recurring maintenance issues are common frustrations for many. Let's dive into these challenges and unpack them a bit.

One of the most significant pain points is undoubtedly the high energy bills. When an HVAC system isn't running efficiently, it can guzzle energy like a thirsty traveler in a desert. This can stem from something as simple as a dirty filter or as complex as a failing compressor.

Either way, the result is higher costs. It's not just about money; it's also about managing the stress and frustration that come with seeing those bills rise month after month.

Recurring maintenance issues are a perpetual thorn in the side. Fixing one problem only to have another pop up soon after can feel like playing an endless game of Whac-A-Mole. It's exhausting and often leads to a sense of helplessness and frustration. But take heart; understanding these challenges is the first step towards conquering them.

Now, let's discuss the financial implications of HVAC system failures or inefficiencies. These aren't just minor inconveniences; they can spell major financial trouble. Inefficiencies in your HVAC system mean more energy usage, leading directly to higher utility bills.

Moreover, HVAC system failures often require costly repairs or, in worst-case scenarios, complete system replacements. These unexpected expenses can throw a wrench into carefully planned budgets, causing financial strain. For homeowners, an unexpected HVAC failure can mean dipping into emergency funds or taking out loans, both of which can have long-term financial repercussions.

The safety risks associated with faulty HVAC systems cannot be overstated. Poor indoor air quality due to neglected systems can lead to health hazards such as respiratory issues, allergies, and long-term illnesses.

Filters need to be changed regularly and ducts must be clean to ensure the air circulating in your home is fresh and free from contaminants. Unmaintained units can also become fire hazards, putting lives and property at risk.

It's understandable if all this information feels overwhelming. Navigating the complexities of HVAC systems can seem daunting. But rest assured, this book will simplify the technical aspects and empower you to make informed decisions. You'll find clear, concise explanations and practical advice that will increase your understanding and confidence in dealing with these systems.

Mastering HVAC fundamentals also taps into our innate desire for independence and self-sufficiency. Isn't it satisfying to fix something with your own hands? Successful DIY projects bring immense satisfaction, and knowing you've contributed to the comfort and functionality of your home is profoundly rewarding.

By learning about HVAC systems, you'll be equipped to tackle small repairs and maintenance tasks yourself, saving money and boosting your confidence. It also makes you a more informed consumer when hiring professionals for more complex jobs, ensuring you get the best service without unnecessary expenses.

Setting the Stage for a Comfortable and Efficient Future

A reliable and well-maintained HVAC system is more than a convenience; it's an essential element of any comfortable, cost-efficient, and safe home. As we saw in Maria's story, neglecting this crucial aspect can swiftly turn a cozy night into a distressing ordeal. Regular maintenance and the use of high-efficiency equipment are fundamental to avoiding such crises.

We previously highlighted how over half of a typical home's energy usage comes from heating and cooling. This underscores not just the importance of maintaining these systems but the significant impact they have on our monthly energy bills.

With rising costs and increasing environmental concerns, homeowners must prioritize energy efficiency. High-efficiency models, when properly installed, can save substantial amounts on energy bills, offering long-term financial relief.

However, the investment in efficient HVAC systems extends beyond monetary savings. It ensures safety, preventing hazardous situations like fires or poor indoor air quality, which can lead to severe health issues. The tragic event involving a faulty air conditioning unit resulting in a deadly fire is a stark reminder of what could go wrong.

Moreover, taking initiative in tasks like changing filters and sealing ductwork plays a critical role in system effectiveness. Up to 20% of conditioned air can be lost through leaky ducts, leading to inefficiencies and discomfort. Homeowners can take simple yet impactful steps to keep their systems running smoothly, ensuring both comfort and cost savings.

From a broader perspective, advancements in HVAC technologies promise even greater efficiency and safety benefits. Projects funded by the Department of Energy aim to develop more innovative solutions, potentially transforming the industry and redefining standards for home comfort.

Think about your own HVAC system—could it benefit from more attention? Taking steps now will ensure you avoid those chilling scenarios and remain confident in the reliability of your home's climate control.

✦ Discover the fundamentals of HVAC systems in this book and test your knowledge at the end to see how well you've mastered the concepts. Keep an eye out for our quiz! ✦

References

Sahoh, B., Kliangkhlao, M., & Kittiphattanabawon, N. (2022). Design and Development of Internet of Things-Driven Fault Detection of Indoor Thermal Comfort: HVAC System Problems Case Study. Sensors (Basel, Switzerland), 22(5), 1925. https://doi.org/10.3390/s22051925

U.S. Department of Energy. (n.d.). HVAC Tech Solutions. Retrieved from https://rpsc.energy.gov/tech-solutions/hvac

U.S. Energy Information Administration. (n.d.). Types and amounts of energy use in U.S. homes and by the types and locations of homes. Retrieved from https://www.eia.gov/energyexplained/use-of-energy/homes.php.

Lundgren-Kownacki, K., Dalholm Hornyanszky, E., Chu, T. A., Alkan Olsson, J., & Becker, P. (2018). Challenges of using air conditioning in an increasingly hot climate. International Journal of Biometeorology, 62(3), 401. https://doi.org/10.1007/s00484-017-1493-z

1
Introduction to HVAC Systems

Have you ever walked into a room where the air feels just right—neither too warm nor too cold, and fresh without being stuffy? Chances are, it's because of an efficient HVAC system working quietly behind the scenes.

While we might take this comfort for granted, understanding the mechanics of HVAC systems opens a world of knowledge that can empower homeowners, DIY enthusiasts, and novice technicians alike.

But achieving and maintaining such indoor comfort isn't as simple as setting your thermostat to a desired temperature. The complexity lies in the coordinated effort of various components like furnaces, air conditioners, ventilation ducts, and thermostats.

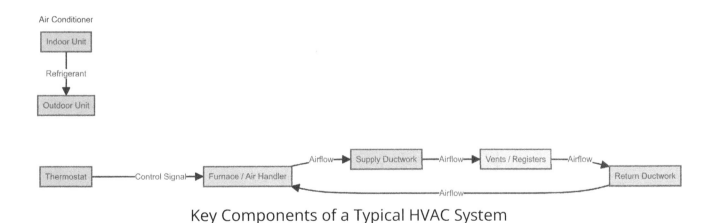

Key Components of a Typical HVAC System

Without essential know-how, even the most common issues can become sources of frustration and unnecessary expenses. Understanding the basics will not only help you recognize minor problems before they escalate but also guide you in effective troubleshooting and maintenance practices.

By demystifying the intricacies of HVAC systems, you'll gain the practical wisdom you need to ensure your home remains an oasis of comfort and efficiency.

Overview of HVAC Components

HVAC systems are like intricate puzzles where each piece plays a critical role in regulating your indoor climate. To get started on this journey of understanding, let's take a closer look at the fundamental components that make up these systems.

HVAC SYSTEM	HEAT SOURCE	DISTRIBUTION METHOD	PROS	CONS
Forced Air	Furnace (Gas, Oil, Electric)	Ducts	Efficient, relatively low installation cost, can be combined with air filtration	Requires ductwork, can be noisy, potential for uneven heating/cooling
Radiant	Boiler (Gas, Oil, Electric) or Heat Pump	In-floor tubing or radiators	Quiet, even heating/cooling, good for allergy sufferers	Higher installation cost (especially for in-floor), slower to respond to temperature changes, not ideal for cooling in humid climates
Geothermal	Ground Heat	In-floor tubing or forced air ducts	Very energy efficient, long lifespan, can provide both heating and cooling	Highest installation cost, requires suitable land for installation, limited to certain climates

Comparison of HVAC System Types

An HVAC system is composed of several key parts: furnaces, air conditioners, ventilation ducts, among others. Each component has its own unique function, yet they must work harmoniously to maintain a stable indoor environment.

Imagine the furnace as the heart of the system, providing warmth during cold seasons. It's typically located in a utility area and connected to a series of supply and return ducts, which serve as the veins and arteries transporting warm air throughout your home.

Just as essential is the air conditioner, the counterpart working tirelessly during hot months to cool down your living space. It's often split into two units: an **indoor** evaporator coil with other accessories and an **outdoor** condenser coil, expansion valve and compressor with other accessories. It should be noted that the location of the expansion valve depends on design.

The evaporator coil absorbs heat from the indoor air, turning it into a gas. This gas then travels to the outdoor unit where it gets compressed and released into the atmosphere through the condenser coil which is also a heat exchanger like the evaporator coil, cooling your home in the process. The heat exchange is made possible through the fan at the outdoor and the blower at the indoor.

When your thermostat signals a need for more warmth or cool air, it sets off a chain of events within the HVAC system. The sequence begins with energy being drawn from the furnace or air conditioner and ends with the conditioned air being pushed through ducts and distributed by vents.

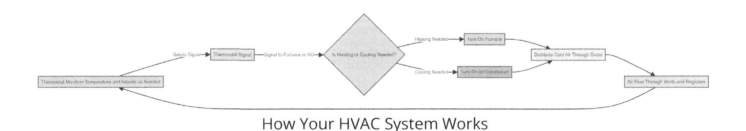

How Your HVAC System Works

Now, imagine encountering a problem with your HVAC system. With essential knowledge about the system's components, you're better equipped to troubleshoot issues. Rather than calling a technician for every minor inconvenience, you could address the problem yourself.

HVAC troubleshooting tips:

- Start by checking if there's power supply to the unit from the mains (that is; if the unit is ON/OFF) and that the thermostat is set correctly and the batteries are functional.

- Ensure the air filters are clean--dirty filters can hinder airflow.

- Look at the condenser coil outside. It might need cleaning if there is debris.

- Examine the blower motor. Is it running smoothly, or making unusual noises?

- Inspect the ductwork for any obstructions or leaks that might affect airflow.

By following these steps, you can diagnose common issues and potentially fix them without professional assistance, saving both time and money.

Furthermore, regular maintenance is vital for the longevity and effectiveness of your HVAC system. Just like you wouldn't skip an oil change for your car, neglecting routine care for your HVAC components can lead to significant problems down the line.

For instance, regularly replacing air filters prevents dust and debris from clogging the system and ensures efficient airflow. Cleaning the evaporator and condenser coils helps maintain their ability to transfer heat effectively. Routine checks on the thermostat and blower motor ensure that they're operating correctly and not causing unnecessary strain on the system.

You can maintain the best possible performance from your HVAC system by keeping a schedule for these tasks:

- Change air filters every one to three months depending on usage and filter type.

- Clean evaporator and condenser coils annually to remove dirt and improve efficiency.

- Check and calibrate the thermostat periodically to ensure accurate temperature readings.

- Inspect blower motors and lubricate their moving parts to prevent wear and tear.

- Examine ductwork yearly for leaks and insulation issues to enhance energy efficiency.

By adhering to these maintenance guidelines, you can prolong the life of your HVAC system and avoid costly repairs in the future. Remember, maintaining your HVAC system doesn't just ensure comfort—it also represents a small but impactful step towards greater individual freedom and social responsibility.

Basic Thermodynamics in HVAC

Understanding the basic principles of HVAC systems is like unlocking the secrets to a comfortable home. It's not just about pressing buttons on a thermostat; it's about understanding the science that keeps your environment cozy in winter and cool in summer.

One of the cornerstone concepts here is thermodynamics, which governs how heat is transferred and managed within HVAC systems.

Thermodynamics at its core is the study of energy transformations and interactions between different systems. Think of it as the blueprint that orchestrates the movement of heat, ensuring that your HVAC system operates efficiently.

According to the first law of thermodynamics, energy cannot be created or destroyed, but can only change forms (PhD, 2017). In an HVAC system, this means that the energy used to heat or cool your home is not lost; it's merely transferred from one place to another.

The second law tells us that heat naturally flows from warmer areas to cooler ones, which our HVAC systems capitalize on to maintain a desired indoor climate.

When you understand these principles, designing an energy-efficient HVAC system becomes far more manageable. For instance, good insulation and well-sealed ducts ensure that most of the energy directed towards heating or cooling benefits your living space, rather than leaking out.

Choosing the right equipment and calculating the correct heating and cooling loads is another vital aspect where thermodynamic knowledge proves invaluable:

- Assess the size and layout of your space. Larger spaces will require more powerful equipment.

- Think about the weather. If you live in a region with extreme temperatures, your HVAC system will need to work harder, so selecting robust, efficient units is crucial.

- Take into account the windows and how they are positioned. South-facing windows might collect more heat during the day, influencing your cooling needs significantly.

- Use load calculation software or consult a professional to determine the precise capacity requirements for your system. Overestimating or underestimating can lead to inefficiency and increased costs.

Diving into the basics of thermodynamics can feel daunting, but it's a worthwhile endeavor. Understanding these principles makes troubleshooting simpler.

If a room isn't reaching the set temperature, it might not just be a mechanical failure but a thermodynamic imbalance—a leaky duct or inadequate insulation causing heat loss. Knowing what to look for can save time, money, and quite a bit of frustration.

And let's not forget the social responsibility aspect. Efficient HVAC systems do more than provide personal comfort; they also reduce energy consumption, contributing to lower carbon emissions. By optimizing your HVAC system based on solid thermodynamic principles, you're making a significant contribution to both your wallet and the planet.

Importance of Proper Ventilation

Among the essential components of HVAC systems, proper ventilation stands out as a cornerstone for ensuring both air quality and overall well-being within any building.

Effective ventilation ensures the circulation of fresh air and removal of unseen pollutants like carbon dioxide, volatile organic compounds (VOCs), and other particulates. These can accumulate in poorly ventilated spaces, causing discomfort and potentially harmful health effects over time.

Think back to those first few months of the COVID-19 pandemic. Public health experts underscored the importance of good indoor ventilation to reduce virus transmission.

Good ventilation doesn't just mitigate viral spread; it also reduces concentrations of dust, allergens, and other airborne irritants that can exacerbate conditions like asthma and allergies.

Here are some things you can do to get and keep good ventilation:

- Open windows and doors, when possible, to allow natural airflow.

- Use exhaust fans and hoods in kitchens and bathrooms to vent smoke, steam, and odors directly outside.

- Ensure that HVAC systems include mechanical ventilation options such as air exchange units to consistently bring in fresh outdoor air and remove indoor contaminants.

- Regularly replace or clean HVAC filters to remove particulates from the air.

- Incorporate air purifiers that use HEPA filters, which can capture 99.97% of particles that are 0.3 microns or larger.

Bad ventilation also allows humidity to build up, which creates a perfect breeding ground for mold and mildew. Mold not only damages building materials but also poses significant health risks, from respiratory issues to severe allergic reactions.

For optimal humidity control:

- Install and regularly maintain dehumidifiers, especially in damp areas like basements, to maintain an optimal range of 30-50%.

- Use programmable thermostats to adjust temperature and humidity levels based on occupancy and weather conditions.

- Ensure that your HVAC system includes proper ductwork with insulated materials to prevent condensation buildup.

Understanding ventilation requirements is vital for compliance with building codes and regulations. Regulations exist for a reason—they provide minimum standards to ensure safety, health, and energy efficiency. These codes dictate everything from minimum air exchange rates to the types of materials permissible in HVAC construction.

For instance, commercial buildings often require more stringent ventilation standards than residences due to higher occupancy rates and different activities performed inside.

Compliance includes:

- Consulting local building codes and standards specific to your region.

- Hiring certified HVAC professionals to assess and install systems according to current regulations.

- Conducting regular inspections and maintenance to ensure ongoing adherence to code requirements.

The key takeaway here is that ventilation is indispensable for creating a comfortable and healthy indoor environment. It's one of those elements that, when properly integrated, quietly works behind the scenes to safeguard our health and enhance our well-being.

Energy Efficiency Considerations

When it comes to HVAC systems, energy efficiency is more than just a buzzword. With advanced technologies and optimized system designs, energy-efficient HVAC systems play a crucial role in managing both expenses and ecological sustainability.

For instance, replacing conventional heating and cooling equipment with high-efficiency alternatives can cut energy use by up to 50% for electric systems and about 10% for gas furnaces. This not only translates to lower utility bills for homeowners, but also reduces the overall demand for energy produced from non-renewable sources, which in turn decreases greenhouse gas emissions.

So, how do these energy-efficient systems achieve such impressive results? Several factors come into play.

Most importantly, these systems often incorporate advanced heat exchangers, variable speed/ variable frequency motors, and smart thermostats for precise control over heating and cooling cycles. These ensure that the system runs optimally, consuming less power while maintaining desired comfort levels.

Implementing energy-saving practices like proper insulation and appropriately sizing your equipment can significantly boost system performance. Let's look at some practical steps you can take to ensure your HVAC system operates as efficiently as possible:

- Ensure your home is well-insulated to minimize heat loss in winter and heat gain in summer. Insulation acts as a barrier, keeping conditioned air inside and reducing the workload on your HVAC system.

- Properly size your HVAC equipment based on a professional assessment. Oversized or undersized systems can lead to inefficiencies and increased wear and tear.

- Install programmable thermostats to manage heating and cooling cycles smartly. These devices can adjust temperatures automatically based on your daily schedule, saving energy when you're asleep or away from home.

It's important to keep up with regular maintenance and timely upgrades in order to improve the energy performance of your HVAC system. Regular maintenance can help catch and address minor issues before they turn into bigger headaches, keeping your system running smoothly.

Here's a simple guideline to follow:

- Schedule a professional inspection n(Planned preventive maintenance) at least twice a year, ideally during spring and fall, to prepare your system for the peak summer and winter seasons.

- Clean or replace air filters monthly. Dirty filters force the system to work harder, leading to higher energy consumption and potential damage to internal components.

- Look into retrofitting older systems with new, energy-saving parts. Upgrading components like ducts, fans, and controls can improve efficiency without the need for a complete system overhaul (YTI Career Institute, n.d.).

Conducting energy audits and assessments is another valuable tool to enhance your energy efficiency. These evaluations help identify areas where your HVAC system may be losing efficiency and provide recommendations for targeted improvements.

For the best results from an energy audit, here are some actions to keep in mind:

- Hire a certified energy auditor to perform a comprehensive evaluation of your HVAC system, including ductwork, insulation, and overall performance.

- Review the audit report to identify specific areas that require attention, such as sealing leaky ducts or upgrading outdated equipment.

- Implement recommended changes gradually, starting with the most critical areas first, to spread out the investment and monitor the impact on energy savings.

It's a win-win situation: you cut down on your energy bills while also playing a part in protecting the environment. With conscious choices and regular maintenance, your HVAC system can serve as a model of efficiency and reliability, keeping your home comfortable for years to come.

You may worry about the complexity and technicality involved in managing HVAC systems. While initial learning might seem daunting, breaking down each task into manageable steps—such as those we've outlined for effective troubleshooting and regular maintenance—can simplify the process significantly.

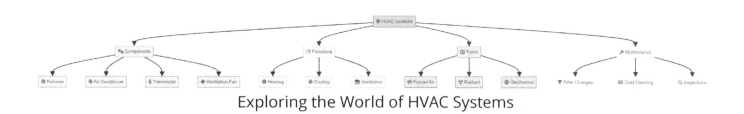

Exploring the World of HVAC Systems

References

Bailes, A. III, PhD. (2017). The Four Laws of Thermodynamics. ACCA HVAC Blog. Retrieved from https://hvac-blog.acca.org/four-laws-thermodynamics/

Florida Technical College. (2021). What are the main components of an HVAC system?. FTC Florida Technical College. Retrieved from https://www.ftccollege.edu/article/what-are-the-main-components-of-an-hvac-system/

Saran, S., Gurjar, M., Baronia, A., Sivapurapu, V., Ghosh, P. S., Raju, G. M., & Maurya, I. (2020). Heating, ventilation and air conditioning (HVAC) in intensive care unit. Critical Care, 24(1), 10.1186/s13054-020-02907-5. https://doi.org/10.1186/s13054-020-02907-5

Harvard Catalyst. (2023). Harvard Professor Highlights the Importance of Ventilation for "Healthy Buildings". Retrieved from https://catalyst.harvard.edu/news/article/harvard-professor-highlights-the-importance-of-ventilation-for-healthy-buildings/

Asim, N., Badiei, M., Mohammad, M., Razali, H., Rajabi, A., Haw, L. C., Ghazali, M. J. (2022). Sustainability of Heating, Ventilation and Air-Conditioning (HVAC) Systems in Buildings—An

Overview. International Journal of Environmental Research and Public Health, 19(2), 1-16. https://doi.org/10.3390/ijerph19021016

RSI. (n.d.). What Are the Principles of Refrigeration and Thermodynamics? Retrieved from https://www.rsi.edu/blog/hvacr/what-are-the-principles-of-refrigeration-and-thermodynamics/

Ross, A. (2023). Components of an HVAC System and How They Work. Erie Institute of Technology. Retrieved from https://erieit.edu/components-of-an-hvac-system-and-how-they-work/

YTI Career Institute. (2024). 10 tips to save energy in HVAC systems. Retrieved from https://yti.edu/blog/10-tips-save-energy-hvac-systems%E2%80%AF

U.S. Department of Energy. (n.d.). Tech Solutions - HVAC. Retrieved from https://rpsc.energy.gov/tech-solutions/hvac

Aresco, R. (2022). Basic Electrical Wiring and Components in HVAC Systems. Erie Institute of Technology. Retrieved from https://erieit.edu/basic-electrical-wiring-and-components-in-hvac-systems/

Bastidas, J. (2023). HVAC-R 101 – What You Need to Know. San Joaquin Valley College. Retrieved from https://www.sjvc.edu/blog/hvac-r-101-what-you-need-to-know/

2

Residential HVAC Installation

Installing residential HVAC systems is a challenging job that demands expertise and attention to detail. It involves carefully considering various factors to achieve optimal efficiency and comfort. When it comes to HVAC installation, it's crucial to ensure that the units are sized and placed correctly.

A small unit that has trouble heating or cooling your space can result in constant operation and higher energy bills. On the other hand, a large unit that turns on and off frequently doesn't do a good job of getting rid of humidity, which can result in an uncomfortable, sticky atmosphere.

Placement is just as important. Improperly positioning units can have negative effects on performance, airflow, and noise levels. Placing the thermostat near windows, drafts or any other position where it will easily come in contact with temperature/condition of the space outside/surrounding can lead to inaccurate readings, which can cause the system to work inefficiently.

Recognizing these possible challenges emphasizes the significance of careful planning for the dimensions and placement of HVAC components. When you follow best practices, you can make sure your HVAC system lasts longer and works better, so your home stays cozy and saves energy.

Importance of Proper Sizing and Placement of HVAC Units

It's crucial to ensure that the HVAC system in a home is properly sized and positioned to guarantee the comfort of its residents. The key lies in understanding how these elements can prevent energy waste and maintain consistent temperatures within your living space.

If you want to figure out the right size for your HVAC unit, it's important to talk to a professional technician. They can do some detailed calculations based on factors like square footage, insulation quality, number of windows, and even the position of your home in relation to the sun (College, 2020). Also, the number of occupants, purpose/usage and heat emitting devices in each particular space are very important considerations.

Here are some things to keep in mind for a consultation:

- Engage a certified technician to assess your home's heating and cooling requirements.

- Ensure the technician conducts a load calculation using industry-standard methods like Manual J from the Air Conditioning Contractors of America (ACCA).

- Confirm that factors such as home insulation, window types, and orientations are included in the assessment.

- Choose an HVAC unit that matches the calculated requirements to avoid undersizing or oversizing.

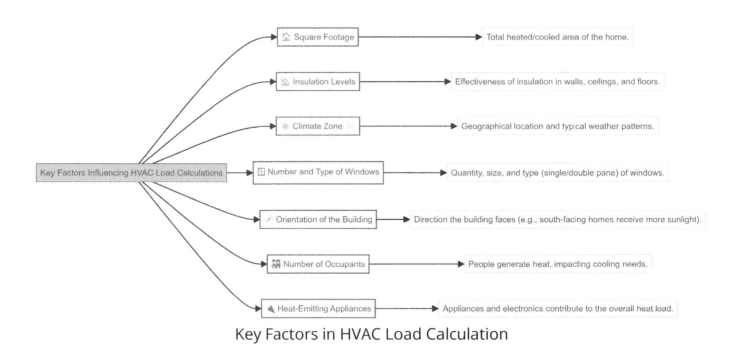

Key Factors in HVAC Load Calculation

Placement considerations involve factors like proximity to windows, air vents, and outdoor elements for optimal performance. The location of both indoor and outdoor components of your HVAC system significantly impacts its efficiency and functionality.

Placing the indoor unit in a central location where the airflow can reach all parts of the home evenly is essential. Ideally, the thermostat should be away from direct sunlight, drafts, and other sources of dust that might affect its reading and thus the system's performance.

For the outdoor unit, it's a good idea to install it in a shaded area to reduce strain during hot weather, but it should also be located away from bedrooms or living spaces to minimize noise disruption.

To make sure the placement is correct, there are a few steps you need to consider:

- Place the outdoor unit in a shady spot and ensure it's easily accessible for maintenance.

- Avoid locations close to bedroom windows to prevent noise disturbances.

- Position the indoor unit centrally to provide even airflow throughout the home.

- Install the thermostat on an inner wall, away from heat-producing appliances and direct sunlight.

Efficient airflow in your HVAC system ensures that every room in your house enjoys a consistent temperature, eliminating any uncomfortable cold or hot spots.

Properly positioned vents and registers are vital in achieving this balance. Make sure there are no obstacles like furniture or heavy draperies that could block the airflow. Think about placing the cooling air outlets higher up on the wall to ensure that cool air is distributed downwards in an efficient manner.

If your system frequently operates outside of its optimal conditions, it can lead to accelerated mechanical wear, which in turn increases the chances of breakdowns and the need for costly repairs or replacements sooner than expected.

To maximize your HVAC system's lifespan and functionality, consider these steps:

- Adhere strictly to manufacturer recommendations for placement and spacing around the unit.

- Schedule regular maintenance checks to ensure all parts are operating smoothly and to catch potential issues early.

- Keep the area around the outdoor unit clear of debris and vegetation to ensure unobstructed airflow.

Consulting with qualified HVAC professionals and following their advice tailored specifically to your home's unique characteristics is the best strategy to ensure your system performs at its best for many years to come.

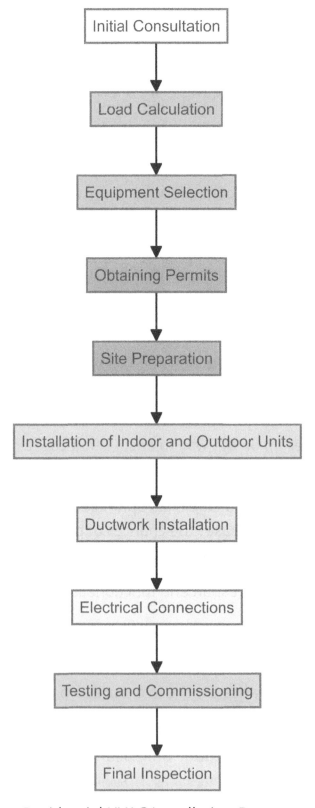

Residential HVAC Installation Process

Intricacies of Ductwork Design

When it comes to ductwork, it's not just about moving air from one place to another: it's about doing so efficiently, quietly, and with minimal energy loss. Let's dive into how you can achieve these goals.

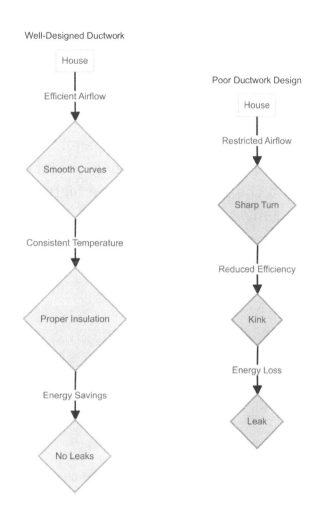

The Impact of Ductwork Design on Airflow and Efficiency

Proper ductwork design ensures efficient airflow, reduces energy consumption, and maintains consistent comfort levels. Here is what you can do to achieve this:

- Start by planning your duct layout carefully to ensure the shortest, most direct path for airflow. This minimizes resistance and keeps your system operating efficiently.

- Use Manual D residential duct design principles (Manual D Residential Duct Systems - ACCA Technical Manuals, n.d.). These guidelines provide a standard method for calculating the size and layout of ductwork based on the specific requirements of your home.

- Select appropriate duct sizes for different parts of the network. Oversized ducts can lead to slower airflow and underused systems, while undersized ones can cause excessive noise and energy use.

- Ensure that your ductwork design includes adequate return air paths. Proper return air flow is crucial for balancing the pressure in your HVAC system, which helps maintain consistent temperatures across different rooms.

Factors like duct material, sizing, insulation, and layout significantly impact the overall performance and effectiveness of the HVAC system.

- For duct materials, options range from sheet metal to flexible plastic or fiberglass. Sheet metal is durable and less prone to sagging, but it can be costlier. Flexible ducts are easier to install and cheaper but may have higher resistance to airflow if not correctly installed (Academy, 2023).

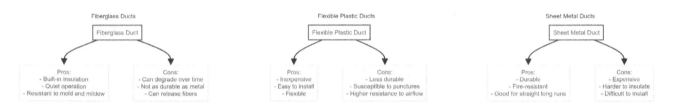

Choosing the Right Duct Material for Your HVAC System

- Proper insulation of ducts is essential, especially in unconditioned spaces like attics or basements. Insulation helps maintain the temperature of the air being moved, leading to better energy efficiency. The insulation also prevents condensation, which can lead to mold growth and other issues.

- The layout of your duct system should avoid sharp turns and right angles as much as possible. These can increase resistance and reduce airflow efficiency. Gentle curves and straight paths are preferable wherever feasible.

- Make sure to account for future changes or upgrades. A modular design allows for easier modifications down the line without major overhauls.

Sealing and insulating ducts properly helps prevent air leaks, ensuring optimal air distribution and minimizing energy loss. Here's how to go about it:

- Use mastic sealant or metal-backed tape to seal all joints and connections in the ductwork. Avoid using standard duct tape, as it loses adhesion over time.

- Inspect both new and existing ductworks for potential leaks, paying particular attention to areas around vents, registers, and seams.

- Install insulation sleeves around ducts in unconditioned spaces. This keeps the air within the ducts at the desired temperature and prevents energy loss.

- Regularly check and maintain the seals and insulation. Age and wear can lead to deterioration, and preventative maintenance can save significant energy costs and improve system efficiency in the long run.

Another key point is the role of return air ducts in promoting good air circulation within the home.

Return air ducts pull air back into the HVAC system to be reheated or cooled (re-conditioned), making sure that air isn't just pumped into rooms but circulated properly. Without a balanced return air system, the supply-side ducts struggle to maintain pressure, causing the entire system to operate inefficiently.

Lastly, think about how ducts might affect the noise level in your home. When it comes to reducing noise from airflow, it's important to focus on proper design and installation.

These factors play a key role in minimizing any unwanted noise. Consider incorporating materials that can absorb sound, like lined ducts or acoustic insulation (silencers), whenever feasible. This enhances the coziness of your living area by ensuring that the HVAC system operates quietly and without causing any disturbance.

It's clear that thoughtful planning and robust design standards can make a considerable difference in both short-term comfort and long-term efficiency. By focusing on these core aspects—efficient airflow, careful consideration of materials, thorough sealing and insulation, and effective ventilation—you can achieve an HVAC system that not only keeps your home comfortable but also operates economically and sustainably.

Installation and Setup of Thermostats

Thermostats are the unsung heroes of our HVAC systems, working tirelessly behind the scenes to regulate heating and cooling cycles based on temperature settings.

At first glance, they might seem like simple devices, but their role in providing comfort while managing energy usage is profound. And while we've come to rely on thermostats for maintaining our ideal indoor climates, there's more than meets the eye when it comes to their installation and setup.

For instance, the location of your thermostat can have a big impact on how accurately and quickly it responds to changes in temperature. This isn't just a matter of sticking it on any convenient wall. Its location can make or break the overall performance of your HVAC system.

Here's what you need to keep in mind about proper thermostat placement:

- Place the thermostat on an interior wall that is central within the home—this allows it to get an accurate reading of the home's average temperature.

- Avoid placing it near drafts, windows, doors, or direct sunlight, as these can skew temperature readings.

- Steer clear of positioning it near ducts or vents since the airflow can interfere with its ability to measure the general room temperature accurately.

Why does this matter?

Accurate temperature readings ensure that your thermostat can effectively communicate with your HVAC system, prompting it to maintain the desired comfort levels efficiently.

In addition to placement, modern thermostats now offer programming features that enable users to easily schedule temperature adjustments. This feature is truly revolutionary when it comes to maximizing energy efficiency and ensuring optimal comfort.

The beauty of programmable thermostats lies in their ability to tailor temperature settings to your daily routines and lifestyle. Here's how you can take full advantage of these features:

- Set different temperatures for weekdays and weekends, aligning with when you're typically home or away.

- Use setback periods—for example, lower the temperature during nighttime hours when everyone is sleeping and crank it back up shortly before you wake up.

- Take advantage of vacation modes that maintain a steady, efficient temperature while you're away, preventing unnecessary energy consumption.

But it doesn't end there. Whether through a mobile app or a control panel, today's tech-savvy thermostats let you personalize your comfort at the tap of a screen. Correct thermostat setup brings an added layer of convenience, allowing you to monitor and adjust indoor temperatures from virtually anywhere.

To set up a smart thermostat correctly, follow these steps:

- Turn off power to your HVAC system to avoid any electrical mishaps.
- Carefully remove the old thermostat and disconnect the wires.
- If required, add a C-wire to provide continuous power to the new thermostat; many modern units need this wire to function optimally.
- Attach the new backplate and reconnect the wires according to the manufacturer's instructions.
- Secure the faceplate and restore power to the system.
- Follow the setup prompts on the thermostat or its companion app to connect it to your Wi-Fi network.

Once set up, you can enjoy the luxury of adjusting your home's temperature while lounging on the couch or even while you're out running errands, ensuring that your living space is always comfortable upon your return.

Thermostats are key to efficient temperature control in residential HVAC systems. They bridge the gap between pure functionality and personalized comfort, helping you maintain the ideal climate in your home with minimal effort.

So, next time you adjust your thermostat, take a moment to appreciate the intricate dance of technology and design in this humble device. By understanding the nuances of thermostat installation and harnessing the power of its advanced features, you're not only enhancing your immediate surroundings but also contributing to a larger narrative of responsible energy use and thoughtful living.

Airflow Optimization Techniques

When airflow is optimized correctly, you can avoid the frustration of experiencing hot and cold spots in different areas. The even distribution of air ensures that every room remains comfortable, which is key to enhancing your overall living experience.

Without proper airflow, certain rooms may become uncomfortably warm while others remain chilly, disrupting both comfort and energy efficiency. This situation forces your HVAC system to work harder than necessary, potentially shortening its lifespan and increasing energy bills.

Maintaining your HVAC system goes beyond just setting it and forgetting it. Regular upkeep of filters and vents plays a significant role in promoting unrestricted airflow, which in turn boosts system performance and indoor air quality.

Here's what you can do to ensure your filters and vents are in top-notch condition:

- Check your air filter at least once a month during peak seasons. If it appears dirty, replace it immediately.

- Keep vents clear of blockages. Ensure furniture or drapes do not obstruct them.

- Schedule periodic professional inspections to clean ducts and remove any accumulated dust and debris.

The benefits of following these tips are twofold: enhanced performance of your HVAC system and a marked improvement in the quality of air within your home (3 Tips to Ensure Optimal Home HVAC Systems Performance This Summer, n.d.).

You should also ensure that the airflow is evenly distributed across different rooms in order to keep a consistent temperature throughout your house and prevent unnecessary strain on your HVAC system. Without proper air distribution, certain areas may experience uneven airflow, resulting in inefficiencies and discomfort.

For optimal airflow, follow these steps:

- Install adjustable vents and louvers that allow you to control the volume of air entering each room.

- Consider investing in a zoning system, which uses multiple thermostats to manage airflow precisely.

- Ensure rooms with doors that are frequently closed have return air pathways to maintain pressure balance.

By implementing these steps, you'll help distribute air more evenly for a more comfortable home environment without overburdening your HVAC system.

Furthermore, using airflow dampers and making necessary adjustments can significantly enhance comfort and energy savings. Airflow dampers are valves or plates that regulate airflow inside a ductwork system.

Properly adjusted dampers can provide customized airflow settings tailored to different preferences and needs within the household. To optimize damper settings:

- Identify all dampers and their current positions; they are usually found within the main duct trunks branching off to individual rooms.

- Adjust the dampers by partially opening or closing them to control airflow to specific areas.

- Monitor temperature changes and adjust further as needed for optimal comfort.

Implementing airflow optimization techniques is essential for maximizing the effectiveness and longevity of residential HVAC systems. It keeps your equipment in good condition, saves you money, and creates a healthier living space. Lastly, it helps decrease the carbon footprint of a household, making a positive contribution to environmental conservation efforts.

References

American Society of Home Inspectors. (n.d.). 3 Tips to Ensure Optimal Home HVAC Systems Performance This Summer. Retrieved from https://www.homeinspector.org/Newsroom/Articles/3-Tips-to-Ensure-Optimal-Home-HVAC-Systems-Performance-This-Summer/16026/Article

Sorry, but the provided HTML content does not contain information that allows for the construction of a citation in APA format.

Erie Institute of Technology. (2023). Step By Step HVAC Installation Guide. Erie Institute of Technology. https://erieit.edu/step-by-step-hvac-installation-guide/

US EPA, OAR. (2014). Heating, Ventilation and Air-Conditioning Systems, Part of Indoor Air Quality Design Tools for Schools. Data and Tools. https://www.epa.gov/iaq-schools/heating-ventilation-and-air-conditioning-systems-part-indoor-air-quality-design-tools

Coyne, C. (2020). Everything You Wanted to Know About HVAC Installation. Coyne College HVAC Installation Guide. Retrieved from https://www.coynecollege.edu/everything-to-know-about-hvac-installation-guide/

Florida Academy. (2023). HVAC Installation Best Practices. Florida Academy. Retrieved from https://florida-academy.edu/hvac-installation-best-practices/

U.S. Environmental Protection Agency. (n.d.). Optimize Airflow and HVAC. https://www.energystar.gov/products/data_center_equipment/optimize-airflow-hvac.

Consumer Reports. (2023). How to Install a Smart Thermostat. Consumer Reports. https://www.consumerreports.org/smart-thermostats/how-to-install-a-smart-thermostat-a1397698896/

Coyne, C. (2020). Everything You Wanted to Know About HVAC Installation. Coyne College. Retrieved from https://www.coynecollege.edu/everything-to-know-about-hvac-installation-guide/

Celsius Marketing. (2022). 5 Common HVAC Airflow Problems. Florida Academy. Retrieved from https://florida-academy.edu/5-common-hvac-airflow-problems/

3

Refrigeration Cycle in HVAC

Ever wondered what keeps the temperature just right in your home, no matter the season?

The answer lies in the intricate processes of HVAC systems--and specifically, the refrigeration cycle stands out as the core mechanism that enables heating and cooling.

Understanding this cycle is like getting a behind-the-scenes look at how comfort is engineered within our living spaces. By delving into the specifics of how refrigerants work within these systems, you'll gain a better grasp of both their complexity and their everyday importance.

One of the key challenges in maintaining an efficient HVAC system is managing the refrigerant cycle properly. For instance, did you know that the transformation of refrigerant from liquid to gas allows it to absorb heat from your indoor air?

This pivotal phase change happens within the evaporator coil. Conversely, when the refrigerant condenses back into a liquid, it releases the absorbed heat outside through the condenser coils.

Problems can arise at any stage of this cycle—whether it's a leak in the refrigerant lines, a malfunctioning compressor, or clogged coils hindering airflow. Each issue can significantly impact the overall performance and energy consumption of your HVAC system, making regular maintenance indispensable.

Overview of Refrigerants

Refrigerants are the lifeblood of HVAC systems. These specialized fluids do their thing by flowing through the system, soaking up heat when it's cool and giving it back when it's hot. It's a pretty cool process, actually. The refrigerant goes through a cycle where it switches between being a liquid and a gas. This helps it transfer heat and keep indoor spaces nice and cozy.

When we talk about cooling, it's all about phase conversion—the transition from liquid to gas. This transformation allows refrigerants to absorb heat effectively, making them indispensable in air conditioning and refrigeration systems.

When the hot air from the indoor space moves across the evaporator coils as a result of conventional air current, it triggers the expansion valve to control the refrigerant flow. This allows the refrigerant to soak up the heat contained in the hot air and change from a liquid to a gas (vapour). The compressor then kicks in, pressurizing the gas, which in turn expels the absorbed heat outside by aid of the outdoor cooling fan (draught mechanism) and through the condenser coil, allowing the cycle to start anew (College, 2020).

The distinction between various refrigerants lies not only in their efficiency, but also in their environmental impact. Over the years, various chemicals have been used as refrigerants, each with its own set of advantages and drawbacks.

Chlorofluorocarbons (CFCs) were once widely used due to their exceptional cooling properties. However, they were found to be highly detrimental to the ozone layer, leading to their ban in 1994. Hydrochlorofluorocarbons (HCFCs), such as R-22 or Freon, became the next best option but are also being phased out due to environmental concerns (Staff, 2023).

More recently, hydrofluorocarbons (HFCs) have emerged as a safer alternative. They do not deplete the ozone layer and offer excellent efficiency, though they still contribute to greenhouse gases.

Among these, R-410A which is a mixture of two refrigerants namely R-125 and R-32 became the standard for HVAC units due to its reliability and lower environmental impact compared to its predecessors. But it's important to note that refrigerants are not interchangeable except the system has been retrofitted. Using the wrong type can damage the system and violate environmental regulations (Murphy, 2016). More recently, newer HVAC systems now run on R-32, which is a more eco-friendly refrigerant compared to R-410A.

It is pertinent to note that refrigerants' impacts to the environment as already stated above are easily classified and monitored using two factors namely the: Ozone Depletion Potential (ODP) and Global Warming Potential (GWP) of the refrigerants, and both are usually stated in numbers.

Proper handling and disposal practices ensure the safety of both the HVAC technician and the environment. Improper management can lead to leaks that not only reduce the efficiency of the system but also harm the ozone layer and contribute to global warming.

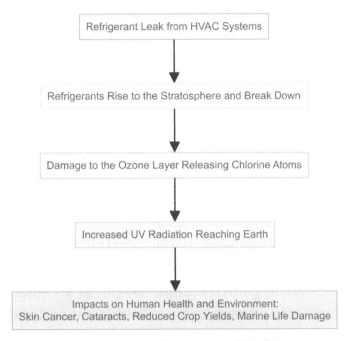

The Environmental Impact of Refrigerant Leaks

Anyone working with HVAC systems needs to adhere to stringent guidelines to prevent environmental contamination:

- Wear appropriate protective gear when handling refrigerants.

- Use certified recovery equipment to capture and recycle refrigerants.

- Follow all local and federal regulations for refrigerant disposal.

- Regularly check for and repair any leaks promptly.

In addition, it's vital to regularly maintain and detect leaks in refrigerant levels to ensure that HVAC systems perform at their best. When refrigerant levels drop, it's usually a sign of a leak. This can cause the system to be less efficient and result in higher energy bills.

Regular inspections are important for catching minor problems before they turn into expensive repairs. Technicians frequently use sophisticated tools such as electronic leak detectors to accurately locate leaks and ensure the system functions seamlessly.

Here is what you can do to maintain your HVAC system effectively:

- Conduct routine inspections every six months to monitor refrigerant levels.

- Use electronic leak detectors for precise identification of leaks.

- Ensure to clean and replace filters regularly to maintain airflow and efficiency.

- Keep outdoor units clear of debris to promote efficient operation.

- Schedule professional maintenance annually to catch and resolve issues early.

Maintaining your HVAC system regularly not only keeps it running efficiently, but also extends its lifespan and minimizes unexpected breakdowns. Furthermore, by following proper maintenance practices, you can keep the system environmentally friendly and reduce the risk of harmful refrigerant leaks.

Additionally, opting for an environmentally sustainable refrigerant not only helps in reducing your carbon footprint but also aligns with the broader goal of preserving our planet for future generations. For instance, switching from R-22 to R-410A in an older unit may require some upfront investment but pays off by ensuring a more efficient and eco-friendly system.

In the end, when it comes to choosing, handling, and maintaining refrigerants, it's not just about following the rules. It's about doing our part to take care of the environment.

Four Stages of the Refrigeration Cycle

The refrigeration cycle is a critical process that keeps our homes and businesses cool. Understanding this cycle can grant you better control over your HVAC system, making maintenance manageable and sometimes even preventing costly repairs.

The essence of the refrigeration cycle is comprised of four main components: the evaporator, compressor, condenser, and expansion valve. These elements work in harmony to transfer heat from an enclosed space to the outside environment, effectively achieving cooling (Inc., n.d.).

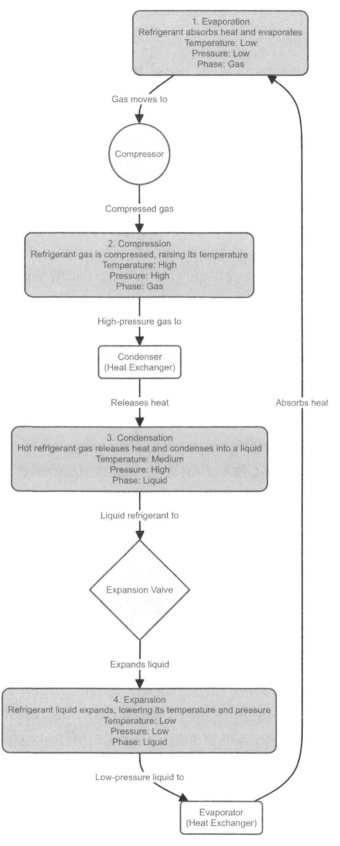

The Four Stages of the Refrigeration Cycle

Think of these components as a well-coordinated team, each playing its part meticulously to keep your surroundings comfortable.

The journey begins with the evaporator. Here, the refrigerant enters as a low-pressure liquid and absorbs heat from the indoor air. This heat absorption causes the refrigerant to evaporate, transforming it into a low-pressure gas filled with indoor heat. It's like a sponge soaking up the warmth from inside your space.

Next, this low-pressure gaseous refrigerant flows to the compressor. The role of the compressor is comparable to a fuel pump in a car—it boosts the pressure of the refrigerant, converting it into a high-pressure, high-temperature gas. By doing so, it prepares the refrigerant for the next stage where it sheds this captured heat. Essentially, the compressor acts as the powerhouse of the cycle, driving the refrigerant forward (Hoffman, 2002-2006).

Following compression, the hot, pressurized gas heads to the condenser. Located often outside the building, the condenser releases the absorbed heat from the refrigerant to the outdoor air, causing the refrigerant to condense back into a liquid state. Imagine turning off a steam iron; when the steam condenses, it turns back into water—this is like what happens within the condenser, albeit at much higher pressures and temperatures (Inc., n.d.).

Finally, the high-pressure liquid refrigerant passes through the expansion valve. This valve throttles the flow of the refrigerant, lowering its pressure dramatically before it re-enters the evaporator. Lowering the pressure results in a drop in temperature, allowing the refrigerant to absorb more heat again as it cycles back to the evaporator. This step is akin to letting air out of a balloon—the abrupt release decreases both pressure and temperature (Hoffman, 2002-2006).

Understanding the order of events in the refrigeration cycle allows you to make informed decisions, so your HVAC system works efficiently. By having this basic understanding, you'll be able to make well-informed choices that balance both comfort and cost-effectiveness.

It's also important to keep in mind that taking care of these components is crucial for keeping your system running smoothly and for a long time. By keeping up with regular maintenance and paying attention to details, you can create a cooler and more comfortable environment. This shows that you can balance economic growth and human welfare, especially in your own home.

Here is what you can do to diagnose and troubleshoot system issues related to refrigeration cycle performance:

- Observe whether the evaporator coil is evenly cold or frosted.

- Listen for any unusual sounds from the compressor indicating wear or damage.
- Check the condenser coils for dirt or blockages that hinder heat release.
- Verify that the refrigerant levels are adequate, and the expansion valve is working correctly.
- Ensure the evaporator and condenser coils are clean and unobstructed.

Proper maintenance also plays a crucial role in ensuring your HVAC system operates efficiently. Regular inspections and servicing can prevent minor issues from escalating into major ones.

Pressure-Temperature Relationship

One essential aspect of this cycle is grasping the pressure-temperature relationship within the system.

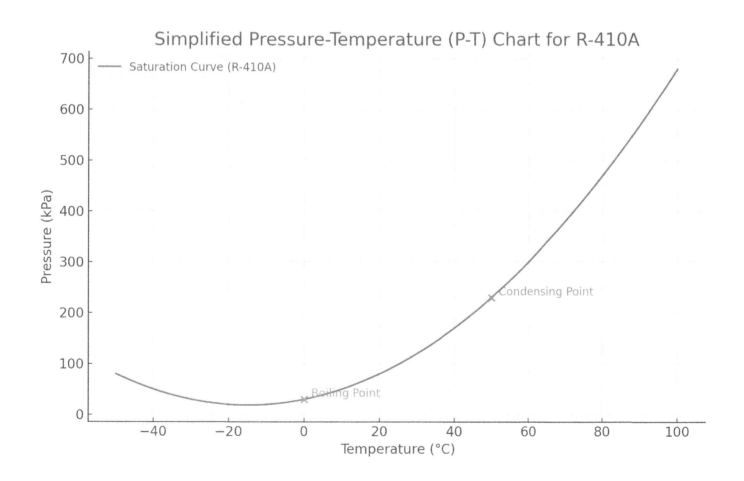

Refrigerants change state—from liquid to gas and vice versa—and in doing so, they absorb or release heat at specific pressures and temperatures. Just as ice takes in heat from its surroundings to melt, refrigerants absorb heat when evaporating. Conversely, when they condense back to liquid, they release that absorbed heat.

Within your HVAC system, this process allows for cooling and heating, based on where and how this change of state occurs—evaporators and condensers play key roles here.

The pressure-temperature chart is an important tool for anyone who works with HVAC systems. These charts show how different refrigerants behave in different situations.

By referencing them, you can determine the current state and performance characteristics of your system:

- Start by identifying the type of refrigerant your system employs.

- Consult the chart corresponding to that refrigerant. You'll see various columns listing pressures and their respective temperatures.

- Match the pressure reading from your system gauges to the chart to find the temperature at which the refrigerant should be evaporating or condensing under ideal conditions.

Following these steps gives you an immediate snapshot of your system's operating efficiency.

When we talk about deviations from optimal pressure-temperature levels, things get interesting—and maybe a little concerning. These deviations can serve as early warning signs of possible system malfunctions or inefficiencies.

If the pressure is too low, your system might have a refrigerant leak or obstruction. If it's too high, it could indicate issues such as overcharging or restrictions in the condenser. Recognizing these signs early on can prevent small problems from escalating into significant repairs.

Maintaining the correct pressure-temperature levels isn't just about avoiding breakdowns; it's also vital for ensuring the reliability and efficiency of the system. Consistent monitoring and regulating can extend the lifespan of your equipment and keep energy costs down. Here's how:

- Regularly check your refrigerant levels and pressures using appropriate gauges.

- Clean and inspect the coils for any signs of dirt or damage.

- Ensure that the airflow through the evaporator and condenser coils is not obstructed.

- Adjust the charge of the system as necessary, either by adding or removing refrigerant to match the manufacturer's specifications.

Understanding the relationship between pressure and temperature is crucial for identifying and solving problems in the refrigeration cycle. If you make mistakes in this area, it could lead to subpar cooling, increased energy costs, and potentially costly repairs.

If your air conditioning isn't cooling effectively, a quick check of the pressure-temperature relationship could reveal the root cause.

Say your system uses R-410A refrigerant. You measure the pressure and find it significantly lower than what the pressure-temperature chart suggests for normal operation. This discrepancy hints that the refrigerant is likely leaking, and you'll need to address it promptly to avoid further complications.

Similarly, suppose your system seems to be running, but it's not cooling the space efficiently. Referencing the pressure-temperature chart, you notice the condenser pressure is unusually high. This observation might lead you to check the condenser coil, which could be dirty or blocked. Cleaning the coil would then help restore normal pressure and improve cooling efficiency.

These guidelines aren't just for experts; they're accessible steps you can take even if you're relatively new to HVAC systems.

The key takeaway? Always go back to the fundamentals—the relationship between pressure and temperature. It remains your best diagnostic tool and your most reliable guide to effective HVAC maintenance.

Common Issues and Troubleshooting

Although the refrigeration cycle is truly the heart of any HVAC system, it can have problems. Knowing what those problems are and how to fix them will help you deal with common problems that may come up and keep your system running smoothly.

Refrigerant leaks are a major concern in HVAC systems. They can lead to reduced cooling capacity, increased energy consumption, and even environmental harm due to the escape of refrigerants into the atmosphere (Energy.gov, n.d.).

Addressing this issue promptly is crucial. Here's what you can do to tackle refrigerant leaks:

- Start by inspecting all visible refrigerant lines for signs of wear, such as oil stains or frosting, which indicate a potential leak.

- Use an electronic leak detector to pinpoint the exact location of the leak.

- Once identified, shut down the system and repair the leak using a suitable sealant or by replacing the damaged components. For split air conditioners, leaks usually occur at connection points called flared ends. Sometimes, leaking copper pipes might require brazing (gas welding). In this case, the attention of a professional technician might be required especially when occupants can't handle such issues.

- Recharge the system with the correct amount of refrigerant, adhering strictly to the manufacturer's specifications. These steps will help you maintain the cooling efficiency of your HVAC system while minimizing its environmental impact.

Another problem that can arise is when the refrigerant levels are not sufficient or when the charging process is done incorrectly. Both situations can have a significant impact on how well the system operates and how effectively it cools. For instance, an overcharged system might overwhelm the compressor, while an undercharged system won't cool properly (Domanski et al., 2016).

Maintaining the right refrigerant levels is key. To manage refrigerant levels effectively:

- Measure the superheat and subcooling values at the evaporator and condenser coils, respectively. These readings will tell you if the system has too much or too little refrigerant.

- Always use manufacturer-approved procedures and tools when adding or removing refrigerant.

- If you're not confident in performing these tasks, hire a qualified technician to avoid potential mishaps. Proper refrigerant management not only boosts performance but also extends the lifespan of your system.

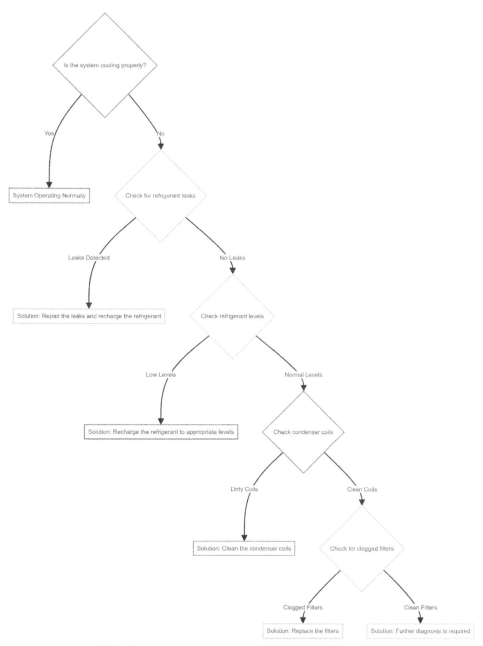

Troubleshooting Common Refrigerant Problems

Dirty condenser coils and clogged filters are often overlooked but they can restrict airflow, hindering heat transfer and affecting cycle performance (ENERGY STAR®, n.d.).

Regular maintenance will prevent such issues. When dealing with dirty condenser coils or clogged filters:

- Begin by turning off the power to your HVAC unit.
- Remove any debris around the outdoor unit to ensure adequate airflow.
- Clean the condenser coils using a coil cleaner and a soft brush; be gentle to avoid

damaging the fins.

- Replace or clean air filters monthly, or as needed based on usage and manufacturer's recommendations. Keeping these components clean will enhance airflow and improve overall system efficiency.

Finally, regular system inspections, leak detection, and cleaning protocols help maintain optimal refrigeration cycle operation. Neglecting these routine checks can lead to severe inefficiencies and costly repairs down the line (Energy.gov, n.d.).

For efficient system maintenance, it's important to follow these guidelines:

- Schedule annual pre-season check-ups with a professional contractor to ensure your system is ready for peak seasons.

- Perform monthly inspections of your system's components, including refrigerant lines, coils, and filters.

- Invest in a smart thermostat, which can help monitor your system's performance and notify you of any irregularities.

- Seal any duct leaks to prevent loss of conditioned air and improve energy efficiency. Adopting these practices will keep your HVAC system running efficiently and prolong its operational life.

As we wrap up this chapter, let's think about how our individual actions are connected to the larger environmental picture. Your choices regarding your HVAC system can have significant consequences. When you stay informed and proactive, you're playing a part in a bigger movement towards sustainability.

TERM	DEFINITION
Refrigerant	A substance used in the refrigeration cycle to absorb heat from one location and release it in another. It undergoes phase changes between liquid and gas states.
Evaporator	The component where the refrigerant absorbs heat from the air or water being cooled, changing from a liquid to a gas (vapor).
Compressor	The component that raises the pressure and temperature of the refrigerant vapor, preparing it for the next stage.
Condenser	The component where the high-pressure, high-temperature refrigerant vapor releases heat to the outside air or water, condensing back into a liquid.
Expansion Valve (or metering device)	The component that reduces the pressure of the liquid refrigerant, causing it to cool and prepare for evaporation.
Phase Change	The transformation of a substance between states of matter, such as solid, liquid, and gas. Refrigerants undergo phase changes during the refrigeration cycle.
Pressure-Temperature Relationship	The direct relationship between the pressure and temperature of a refrigerant. Higher pressure corresponds to higher temperature, and vice versa.
Superheat	The difference between the actual temperature of the refrigerant vapor and its saturation temperature at a given pressure. It ensures that only vapor enters the compressor.
Subcooling	The difference between the actual temperature of the liquid refrigerant and its saturation temperature at a given pressure. It ensures that only liquid enters the expansion valve.

Key Terms in the HVAC Refrigeration Cycle

References

Murphy, D. (2023). Types of refrigerant in your HVAC services. ITI Technical College. Retrieved from https://iticollege.edu/blog/types-of-refrigerant-in-your-hvac-training-service-training/

RSI. (n.d.). What is Refrigerant in an HVAC System? Retrieved from https://www.rsi.edu/blog/hvacr/what-is-refrigerant-in-an-hvac-system/

U.S. Environmental Protection Agency. (n.d.). How to Keep Your HVAC System Working Efficiently. Retrieved from https://www.energystar.gov/products/ask-the-experts/how-keep-your-hvac-system-working-efficiently

Du, Z., Domanski, P. A., & Payne, W. V. (2016). Effect of common faults on the performance of different types of vapor compression systems. Applied Thermal Engineering, 98(61), 10.1016/j.applthermaleng.2015.11.108. https://doi.org/10.1016/j.applthermaleng.2015.11.108

RSI. (n.d.). Four types of refrigeration systems you need to know. RSI. Retrieved from https://www.rsi.edu/blog/hvacr/four-types-refrigeration-systems-need-know/

Southwest Wisconsin Technical College. (n.d.). Basic cycle of air conditioning. Retrieved from https://www.swtc.edu/Ag_Power/air_conditioning/lecture/basic_cycle.htm

He, P., Liang, Y., Hu, Y., Zhang, C., Zhang, D., Ai, X., Wang, Y., & Weng, Y. (2021). Effect of physical properties of a gas on the refrigeration temperature drop of vortex tubes used in oil and gas fields. ACS Omega, 47(6), 31738. https://doi.org/10.1021/acsomega.1c04421

U.S. Department of Energy. (n.d.). Common Air Conditioner Problems. Retrieved from https://www.energy.gov/energysaver/common-air-conditioner-problems

Summit College. (2020). How Refrigerants Work in the HVAC System. Summit College. Retrieved from https://summitcollege.edu/how-refrigerants-work-in-the-hvac-system/

4

Heat Transfer in HVAC

Understanding how heat transfer works is at the core of efficient HVAC operation. It's all about how warmth or cooling is distributed throughout a space.

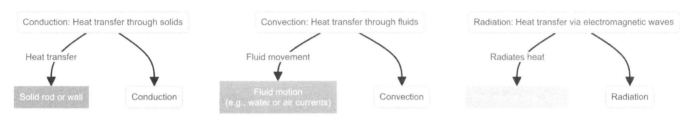

Modes of Heat Transfer in HVAC Systems

Heat transfer can be complex, predominantly occurring through three mechanisms: conduction, convection, and radiation. Heat transfer in HVAC systems affects how heat moves through walls, pipes, and even the parts that make up the machines. Mismanagement of heat flow can lead to inefficiencies such as unwanted heat loss in winter or gain in summer. It's important to choose materials for HVAC parts that have the right thermal values.

Understanding Modes of Heat Transfer in HVAC Systems

In the world of HVAC systems, understanding the different modes of heat transfer is essential for designing systems that work efficiently and maintaining comfort in our homes. Let's explore these principles to get a clearer picture of how heat moves through these systems.

Conduction

Conduction involves the transfer of heat through direct contact between materials. This process can be thought of as heat traveling through a solid medium.

For instance, when you touch a hot stove, you directly feel the heat from the stove transferring to your hand. In HVAC systems, conduction plays a crucial role in understanding how heat flows through solid components like walls, ductwork, and even within the machinery itself.

When designing or modifying an HVAC system, it's vital to consider the materials used and their thermal conductivity. A material with high thermal conductivity will transfer heat more readily, which can either be a benefit or a challenge depending on the design goals.

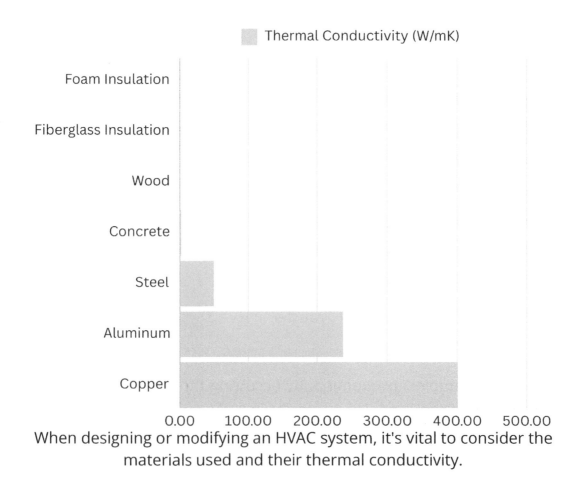

When designing or modifying an HVAC system, it's vital to consider the materials used and their thermal conductivity.

So, ensuring that components prone to significant heat loss or gain are well-managed can make your HVAC system more efficient.

Convection

Convection, on the other hand, relates to the transfer of heat through the movement of fluids or air. As the water at the bottom of a pot heats up on the stove, it becomes less dense and rises, while the cooler, denser water sinks to replace it. This cycle creates a flow that distributes heat throughout the pot.

Similarly, in HVAC systems, convection is harnessed to distribute heated or cooled air efficiently throughout a space. The air inside your home is continuously moving, thanks to your HVAC system's fans and blowers.

These components ensure that warm air circulates during winter and cool air during summer, maintaining a comfortable indoor environment. By optimizing this airflow, you can greatly enhance your system's efficiency and effectiveness.

Radiation

Radiation is another mode of heat transfer and occurs through electromagnetic waves. Picture standing in front of a fireplace; you can feel the warmth without touching the flames. That's radiant heat at work.

HVAC systems canuse radiation in designs that emit or absorb radiant heat. Radiant floor heating systems use pipes buried beneath the floor to radiate heat upward, warming the room efficiently. Understanding the role of radiation in HVAC allows us to create systems that can manage energy better and provide consistent comfort levels within our homes.

Effective insulation methods are critical in minimizing heat transfer losses and improving energy efficiency in HVAC systems. Insulation works by creating barriers to conductive, convective, and sometimes radiative heat flows.

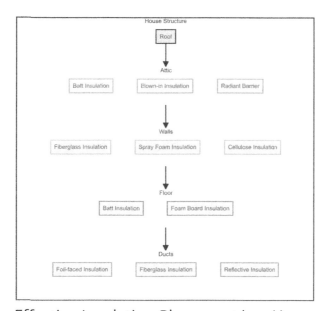

Effective Insulation Placement in a Home

To achieve optimal insulation, here are some guidelines:

- Identify areas in your home where insulation is most needed, such as attics, walls, floors, and ductwork.

- Select the appropriate type of insulation material based on its R-value, which indicates resistance to heat flow. Materials with higher R-values offer better insulation.

- Ensure proper installation of insulation to avoid gaps or compression that can reduce its effectiveness. Pay special attention to sealing corners, edges, and joints.

- Consider additional measures like weatherstripping for doors and windows to prevent air leaks, further enhancing your home's overall thermal efficiency.

Using these steps, you can ensure that your homekeeps warm during winter and stays cool during summer, significantly reducing the load on HVAC systems and cutting down energy costs.

Understanding these mechanisms of heat transfer—conduction, convection, radiation, and effective insulation—is critical in designing HVAC systems for optimal performance and energy efficiency. Recognizing how heat moves and interacts within different mediums allows you to ensure that your HVAC systems are not only effective in regulating indoor temperatures but also efficient in terms of energy consumption.

Role of Heat Exchangers in HVAC Systems

Heat exchangers enable HVAC systems to efficiently transfer heat between separate materials. This feature is essential for optimizing energy usage and maintaining a comfortable internal temperature.

In an HVAC system, a heat exchanger typically works by taking thermal energy from one fluid stream and transferring it to another. Take, for example, your common air-handling unit: as warm indoor air flows over a cooler surface, heat gets transferred through conduction, thus cooling the air before it's recirculated back into the space.

The beauty of this process is that you can manipulate indoor climates effectively without mixing the two fluids directly. This not only saves energy but also enhances air quality by avoiding cross-contamination of different fluid streams (Nguyen, 2018).

The applications of heat exchangers in HVAC extend beyond mere temperature regulation. They're vital components in systems designed for air handling, refrigeration, and even water heating.

When integrated into an air-handling unit, for instance, heat exchangers aid in recovering waste heat, which would otherwise be lost to the environment. By capturing and reusing this heat, these units become more efficient, reducing overall operational costs and decreasing the energy footprint of a building.

In refrigeration systems, heat exchangers play an essential role in both condensing and evaporating phases of refrigerants. This process allows the system to take in heat from one location (such as the inside of your refrigerator) and then release it into another location (typically outside). This guarantees that perishable foods are maintained at the ideal temperatures, all while avoiding expensive energy costs.

Additionally, water heating systems benefit from heat exchangers by harnessing waste heat to warm up water supplies, making these systems more sustainable and cost-effective.

Different types of heat exchangers address various HVAC requirements.

- Plate heat exchangers consist of multiple thin plates that are stacked together, enabling efficient heat transfer in a small area. These tools come in handy when you're dealing with tight spaces and need to maintain optimal performance.

- Shell and tube heat exchangers are composed of a series of tubes enclosed within a shell. These are perfect for situations where you need to transfer a lot of heat under high-pressure conditions. These HVAC systems are perfect for industrial settings or larger commercial applications due to their strong construction.

- Finned tube heat exchangers add another layer of complexity and efficiency. When fins are added to the outside of tubes, it increases the surface area for heat transfer, making these exchangers more effective. They are commonly used in air conditioning units and other situations where quick heat dissipation is needed.

To maintain peak performance, regular upkeep of heat exchangers is key. Here's what you can do to ensure your heat exchangers remain in top condition:

- Begin by powering off the HVAC unit to avoid any electrical hazards.

- Inspect filters routinely, cleaning or replacing them every one to three months as recommended by the manufacturer. Dirty filters impede airflow and diminish the system's efficiency.

- Examine the outside air intakes to make sure nothing is obstructing the screens and hoods.

- Check the condensation pan and drain tubing. Pour some water into the pan to ensure proper drainage; if it doesn't drain, the tubing likely needs cleaning.

- Make sure to clean the heat exchanger core every year, following the manufacturer's guidelines. Remember to keep the power off while doing this. Additionally, make sure to clean the fans and carefully wipe the blades. If the manufacturer specifies, you can also lubricate the motor.

Understanding and maintaining these devices can go a long way toward ensuring that HVAC systems run smoothly and efficiently. Proper maintenance of heat exchangers not only improves their performance, but also increases the lifespan of the entire HVAC system. This leads to better service and lower energy costs.

Importance of Insulation Techniques in HVAC Systems

When it comes to enhancing energy efficiency in HVAC systems, insulation plays a critical role in reducing heat loss and gain, which can significantly impact both the performance and energy consumption of your system.

Types of insulation: While there are various insulation materials available for use in HVAC systems, each serves to reduce either heat loss or gain effectively. Fiberglass, foam, and cellulose are common choices.

Each material has its own set of advantages. Fiberglass is cost-effective and widely used. Foam offers high insulating value per inch. Cellulose is made from recycled paper products, making it an eco-friendly option. The choice depends on your specific needs and budget constraints.

Insulation placement: Proper placement of insulation is just as crucial as selecting the right material. Ensuring that ductwork, walls, and pipes are adequately insulated can prevent significant energy wastage and help maintain desired indoor temperatures efficiently.

Here's what you can do to ensure correct insulation placement:

- Identify all areas where ducts run through unconditioned spaces, such as attics and basements.

- Choose an appropriate type of insulation for these areas, considering factors like moisture levels and accessibility.

- Thoroughly wrap the ducts with the chosen insulation material, ensuring there are no gaps.

- Use approved sealing materials to fix any joints or seams in the insulation, providing an airtight barrier.

Thermal comfort benefits: Effective insulation does more than conserve energy—it enhances thermal comfort within your living space by minimizing temperature variations. Imagine walking through your home and experiencing consistent temperatures in every room, instead of pockets of hot and cold air.

This stability not only makes the environment more pleasant but also reduces the strain on your HVAC system, leading to prolonged equipment life and lower maintenance costs.

Environmental impact: Embracing sustainable insulation materials can take your HVAC system's environmental friendliness to another level. For instance, choosing products with high recycled content or those that have lower embodied energy can substantially reduce your carbon footprint.

By doing so, you're not only conserving energy but also contributing to a more sustainable planet. Additionally, opting for materials that have fewer chemicals prevents indoor pollution, safeguarding your family's health while being kind to the environment.

The effectiveness of insulation is often measured in terms of R-value, where higher values signify better resistance to heat flow. When choosing materials for your home, it's important to consider the R-values that are suitable for your climate zone and specific application areas. Attic insulation may need a higher R-value than wall insulation because of the varying dynamics of heat loss.

Moreover, it's worth noting that well-insulated homes not only offer cost savings but also promote safety and better air quality. According to ENERGY STAR® (n.d.), properly sealed and insulated ducts can avoid "back drafting" of combustion gases such as carbon monoxide, ensuring these harmful gases are expelled outside rather than leaking back into your living spaces. By keeping out pollutants, this helps to minimize risks associated with gas appliances and improves the overall indoor air quality.

Switching gears slightly, let's consider some practical steps you can take to improve insulation in your home.

Start with a thorough inspection to identify under-insulated areas; use thermal imaging cameras if necessary. You can then add insulation to your attic, basement, and exterior walls. Ensure that cavity walls are filled without compressing the insulation material, as this could diminish its effectiveness.

Also, think about how air sealing can complement insulation and provide additional benefits. Gaps, cracks, and other leaks can undermine even the best insulation efforts. Use caulking and weather-stripping to seal these small openings, especially around windows and doors. This dual approach can vastly improve your home's energy efficiency and comfort.

Thermal Comfort Considerations in HVAC Design

Your home should be more than just four walls and a roof; it should be a sanctuary where you feel comfortable, safe, and healthy. Optimizing your indoor environment with proper HVAC design can make all the difference.

Humidity is a key factor. Proper humidity levels are more than just a matter of comfort—they're vital for your health. Both high and low humidity can lead to a host of problems. Too much moisture can lead to mold growth, while dry air can cause respiratory problems and harm wooden furniture and flooring.

Modern HVAC systems often come equipped with dehumidification and humidification features to help strike the right balance.Here is what you can do to achieve optimal humidity levels:

- Invest in a hygrometer to monitor indoor humidity levels regularly.

- Consider incorporating a whole-house humidifier or dehumidifier to maintain ideal conditions, typically between 30% and 50% relative humidity.

- Check and replace filters regularly to ensure they function effectively.

- Ensure that your HVAC system is properly sized for your home to avoid overworking the humidification features.

Efficient airflow distribution and proper zoning are essential to maintaining uniform temperatures throughout different areas of your home. Systems that are not well-designed can lead to inconsistent heating and cooling, causing certain rooms to become excessively hot while others stay uncomfortably cold.

By implementing a well-thought-out zoning strategy, you can direct airflow to different parts of your home based on specific needs, consequently enhancing overall comfort.

To achieve effective zoning and airflow:

- Start by assessing the layout of your home and identifying zones with varying heating and cooling requirements.

- Use dampers in ductwork to manage airflow to distinct sections of the home.

- Install programmable thermostats in each zone for precise temperature control.

- Conduct regular maintenance to ensure fans and ducts are clean and unobstructed for optimal performance.

Energy-efficient HVAC practices not only enhance thermal comfort but also contribute to reducing your home's carbon footprint. By incorporating smart HVAC controls and energy management systems, you can achieve maximum efficiency while still maintaining optimal comfort levels.

Smart thermostats, for instance, learn your habits and adjust the temperature accordingly, optimizing both comfort and energy usage.Here are some tips for boosting energy efficiency:

- Set your thermostat to adjust automatically based on your schedule to reduce energy use when you're not home.

- Use ceiling fans to aid in air circulation, so you can set your thermostat a few degrees higher in summer and lower in winter without sacrificing comfort.

- Seal leaks and add insulation to minimize loss of conditioned air.

- Schedule regular professional service check-ups to keep your system running efficiently.

The quality of your indoor air is another critical component of thermal comfort and overall well-being. Poor air quality can aggravate allergies, complicate asthma symptoms, and foster an unhealthy living environment.

An effective HVAC system should not only regulate temperature and humidity but also filter pollutants and manage ventilation to ensure clean, fresh air circulates throughout your home.To improve indoor air quality:

- Fit your HVAC system with high-efficiency particulate air (HEPA) filters, which capture fine particles like pollen, dust mites, and pet dander.

- Use ultraviolet (UV) germicidal lights to kill bacteria and mold spores within the HVAC system.

- Integrate mechanical ventilation systems to introduce outdoor air, diluting indoor pollutants.

- Conduct air quality tests periodically to identify and address potential issues promptly.

Understanding and applying these principles doesn't require you to be an HVAC expert—just a homeowner dedicated to improving your living space.

References

Kachhava, P., Wate, S., & Ghosh, S. S. (2021). Machine learning-based forecasting of electricity demand in the era of Industry 4.0. Energy Reports, 7, 1578-1588. https://doi.org/10.1016/j.egyr.2021.06.045

Dean, K., & Sharma, R. (2017). Air-to-air heat exchangers for healthier energy-efficient homes. NDSU Extension. Retrieved from https://www.ag.ndsu.edu/publicationS/energy/air-to-air-heat-exchangers-for-healthier-energy-efficient-homes

RSI. (n.d.). Heat exchanger for HVAC systems. RSI Blog. Retrieved from https://www.rsi.edu/blog/hvacr/heat-exchanger-hvac-system/

Phys.org. (2019). New approach could make HVAC heat exchangers five times more efficient. Phys.org. https://phys.org/news/2019-08-approach-hvac-exchangers-efficient.html

Energy.gov. (n.d.). Insulation saves homeowners money and improves comfort. https://www.energy.gov/energysaver/insulation

Energy.gov. (n.d.). Principles of Heating and Cooling. Retrieved from https://www.energy.gov/energysaver/principles-heating-and-cooling

U.S. Environmental Protection Agency. (n.d.). Benefits of Duct Sealing. Retrieved from https://www.energystar.gov/saveathome/heating-cooling/duct-sealing/benefits

HVAC Global. (2022). Types of Heat Transfer. Retrieved from https://hvacglobal.org/2022/01/27/types-of-heat-transfer/

RSI. (n.d.). Making the Most of Your Heat Exchanger in Your HVAC System. RSI. Retrieved from https://www.rsi.edu/blog/hvacr/heat-exchanger-hvac-system/

5

Troubleshooting HVAC Systems

Maintaining the efficiency of your HVAC system often feels like cracking a code, but with the right techniques and tools, you can navigate its complexities with confidence. This chapter lights the path forward by offering insights into the essential diagnostic tools that make tackling these challenges not only feasible but also methodical.

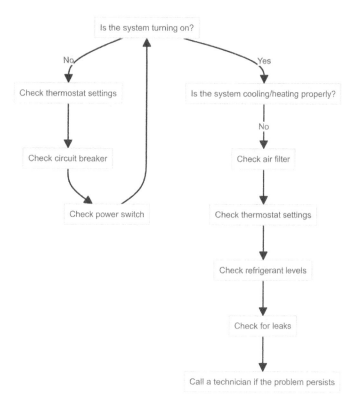

Basic HVAC Troubleshooting Flowchart

When it comes to HVAC systems, diagnosing issues accurately is half the battle won. Diagnostic tools such as multimeters, pressure gauges, digital manifolds, and thermal imaging cameras become indispensable here, serving as your eyes and ears to uncover hidden faults and inefficiencies.

Importance of Diagnostic Tools

Let's dig into some key diagnostic tools that will aid in your HVAC troubleshooting journey.

Diagnostic tools such as multimeters and pressure gauges are essential for measuring electrical and pressure parameters in your HVAC system. They provide a starting point for accurate measurements. These instruments have the ability to detect faulty components that are not visible to the naked eye.

Using a Multimeter to Test HVAC Electrical Components

For instance, a multimeter can measure voltage, current, and resistance, helping you detect electrical issues such as short circuits or broken wires. Similarly, pressure gauges are essential for evaluating refrigerant levels, ensuring that they're neither too high nor too low, which could lead to system inefficiencies or failures.

Here is what you can do to use these tools effectively:

- **First**, always read the manual of the diagnostic tool. Understanding its functions and limitations will give you the confidence to use it correctly.

- **Second**, prior to taking any measurements, ensure the HVAC system is turned off to avoid any electrical hazards.

- **Third**, connect the multimeter probes to the corresponding terminals and select the appropriate measurement type (voltage, current, resistance). Record the readings and compare them against standard values to determine if there's an issue.

- **Fourth**, for pressure gauges, attach the gauge to the service port. Note the pressure level displayed and refer to the manufacturer's specifications to see if it falls within the acceptable range.

Digital manifolds take HVAC diagnostics to another level by displaying real-time data, which allows for quick and informed decision-making. Digital manifolds provide accurate readings and often include extra features such as data logging and error alerts, making them superior to traditional analog gauges. This allows technicians and DIY enthusiasts to speed up their diagnostics and minimize downtime.

Refer to this guideline for using digital manifolds:

- Begin by turning on the digital manifold and familiarizing yourself with its interface. Most units have user-friendly menus and display screens.

- Connect the hoses of the digital manifold to the corresponding service ports on the HVAC unit. Ensure the connections are tight to prevent leaks.

- Monitor the readings displayed on the screen. Look for anomalies or deviations from normal operating conditions. Some digital manifolds offer diagnostic suggestions based on the data collected, making it easier to identify specific problems.

- Save or record the data for future reference. Many digital manifolds allow you to export data to a computer or smartphone for detailed analysis.

Thermal imaging cameras offer a visual method of detecting temperature variations, indicating potential issues like blocked vents or insulation gaps. They are especially useful for spotting problems that are not immediately obvious, such as heat loss through poorly insulated areas or blockages that restrict airflow.

To effectively use thermal imaging cameras:

- **First**, power on the camera and let it calibrate. This ensures that the temperature readings are accurate.

- **Next**, scan the area around the HVAC system, paying close attention to ductwork, vents, and insulation. Look for unusual hot or cold spots that deviate from the norm.

- **Then**, document these areas of concern with snapshots. Many thermal cameras allow

- **Lastly**, use the captured images to guide your physical inspection. Areas identified by the thermal camera should be inspected for possible obstructions or insulation deficiencies and rectified accordingly.

It's really important to regularly calibrate and maintain your diagnostic tools so that you can get accurate readings and reliable troubleshooting results. Over time, even the best tools can deviate from their initial settings, resulting in inaccurate diagnoses. Calibration is the process of making adjustments to a tool using established standards in order to restore it to its proper state.

Maintenance, on the other hand, includes cleaning, battery replacement, and checking for wear and tear.Follow these teps for maintaining and calibrating your diagnostic tools:

- Inspect the tools regularly for signs of damage or wear. Replace parts as needed to maintain operational integrity.

- Clean the tools periodically. Dust, grime, and moisture can affect the performance of sensitive diagnostic equipment.

- Calibrate the devices according to the manufacturer's instructions. Use calibration standards (like a known voltage source for a multimeter) to adjust the readings. Professional calibration services are available if the tools require complex adjustments.

- Store the tools properly. Use protective cases and keep them in controlled environments to prevent damage from environmental factors like humidity and extreme temperatures.

Always remember that while personal responsibility plays a significant role in maintaining your HVAC system, having a reliable safety net of well-calibrated and maintained diagnostic tools is indispensable.

Addressing Airflow Restrictions

When filters are clogged with dust, pollen, and other airborne particles, they restrict airflow, forcing your system to work harder than necessary (Marketing, 2022). This increased effort not only consumes more energy but also reduces the system's efficiency, leading to uneven cooling or heating and potential damage over time.

Scheduling regular filter checks and replacements is a small yet pivotal task that can avert significant issues, such as skyrocketing energy bills or premature system wear-out. Here's what you can do to stay ahead:

- Regularly inspect your air filters, ideally every three months, though more frequent checks might be necessary in dusty environments or homes with pets.

- Always have replacement filters on hand to avoid delays when it's time for a change.

- Consider upgrading to higher-quality filters if your current ones consistently become clogged quickly.

- Set reminders on your phone or calendar to ensure these checks become part of your routine maintenance.

Inspecting ductwork for obstructions, leaks, or improper sealing is another critical step in maintaining optimal airflow. Airflow restrictions can sometimes be tricky to detect as they can lurk within the hidden maze of ducts, only becoming apparent once they have severely affected the performance of the system. Even minor leaks or misalignments can cause pressure imbalances that can affect the efficiency and effectiveness of your HVAC system.

Over time, gaps or cracks in the ductwork can cause heated or cooled air to escape into unconditioned spaces, such as attics or crawl spaces, leading to reduced comfort in living areas and increased utility bills.To counteract these issues:

- Conduct a visual inspection of accessible ducts for obvious signs of damage or disconnection.

- Use a smoke pencil or incense stick to identify less apparent leaks; hold it near duct seams and observe any disturbances in the smoke flow.

- Seal minor leaks with mastic sealant or metal tape rather than duct tape, which tends to degrade over time.

- If you come across substantial damage or inaccessible sections, consider enlisting a professional for a more thorough inspection and repair.

Using airflow meters can dramatically enhance your ability to measure air velocity and volume, thus aiding in diagnosing restrictions and balancing airflow distribution. These tools provide quantitative data, offering a clear picture of how air travels through your system.

With an airflow meter, you can pinpoint trouble spots that may be causing inefficiencies, such as blockages or poorly designed duct layouts, and take corrective actions accordingly. For effective use of airflow meters:

- First, refer to your HVAC system's manual to understand the optimum airflow levels required for different components.

- Begin by measuring the airflow at the supply and return vents to get a baseline reading, then proceed to check various points along the ductwork.

- Compare these readings against manufacturer specifications and note any discrepancies.

- Adjust airflow dampers or balance dampers as needed to even out the distribution throughout your home.

- Document your findings and recheck periodically to monitor changes and maintain system balance.

By implementing zoning strategies and making damper adjustments, you can have greater control over how air is distributed in various parts of your building. This enables you to fine-tune the airflow to meet specific needs in each area.

Zoning essentially involves creating separate sections within the HVAC system, each controlled independently to cater to specific needs or preferences. This helps manage temperature variations across multiple rooms, enhancing overall comfort while optimizing energy usage. Here's how you can approach zoning:

- Identify areas with distinct heating or cooling requirements, such as occupied versus unoccupied rooms, or areas with more sun exposure.

- Install zone dampers in the ductwork and connect them to individual thermostats within each zone. These dampers adjust based on the thermostat settings, directing air where it's most needed.

- Program the thermostats to match daily occupancy patterns, ensuring each zone gets the right amount of conditioned air at the right times.

- Regularly test the damper positions and thermostat responses to confirm they're working as intended, adjusting as necessary.

Proactive management of airflow restrictions ensures consistent system operation and improves overall HVAC performance. By addressing these common issues, homeowners and novice technicians alike can enjoy a more comfortable indoor environment and potentially extend the lifespan of their HVAC systems.

Taking care of your HVAC system doesn't have to be daunting. It's about developing a routine and paying attention to the nuances of your system. Regular checks and proactive measures can prevent small issues from snowballing into costly repairs, ultimately safeguarding both your comfort and your wallet.

Remember, knowledge and timely action are your best allies in maintaining a healthy, efficient HVAC system.

Handling Refrigerant Leaks

Dealing with refrigerant leaks and recharges is not just a technical task; it's an essential procedure that safeguards HVAC systems, enhances efficiency, and protects our environment. Let's walk through the steps you'll need to troubleshoot these issues effectively.

Detecting refrigerant leaks is the first crucial step. A common method is UV dye testing. This involves adding a small amount of dye to the system, running it for a short period, and then inspecting with a UV light to spot any dye that has escaped.

On the other hand, electronic leak detectors are incredibly efficient and can detect even the tiniest leaks by recognizing halogen gases that are often present in refrigerants. Both methods work well, but whichever you choose, early detection is paramount to prevent both system inefficiency and potential environmental harm.

Here's what you can do to achieve this:

- Ensure your HVAC system is turned off and cool before starting the inspection.

- For UV dye testing, inject the dye into the system according to the manufacturer's instructions and run the system for enough time to allow the dye to circulate.

- Use a UV lamp to inspect all joints, connections, and components of the HVAC system. Look for bright yellow-green areas indicating a leak.

- For electronic detectors, turn on the device and pass the probe close to possible leak points on the system—connections, valves, and seams are typical spots to check.

Properly detecting leaks not only maintains the operational efficiency of your system but also contributes to environmental preservation by preventing harmful refrigerants from escaping into the atmosphere.

Understanding and adhering to EPA regulations for refrigerant handling is your next step. These rules aren't just bureaucratic red tape; they're fundamental to protecting the environment and ensuring your system runs efficiently.

According to Section 608 of the Clean Air Act, technicians who service HVAC systems must follow specific guidelines to maximize the recovery and recycling of refrigerants, which can be ozone-depleting substances or potent greenhouse gases (EPA et al., 2015).

Here's what you can do to stay compliant:

- Familiarize yourself with the EPA's general leak repair requirements. For instance, if your system contains more than 50 pounds of refrigerant, any detected leak exceeding the trigger rate (10% for comfort cooling appliances) must be repaired within 30 days (EPA et al., 2015).

- Regularly check for updates on regulations as these can change. The EPA frequently revises rules to incorporate new findings and technologies.

- Ensure that anyone servicing your HVAC system is properly certified. They should be able to demonstrate compliance with these standards to avoid penalties and ensure safety.

Prompt action in performing leak repairs is crucial too. If a leak is detected, it needs sealing immediately or, in more severe cases, you may need to replace the faulty component. If leaks are not addressed, they can cause a significant loss of refrigerant, which in turn reduces the efficiency of the system and increases the risk of system failures.

Here's what you can do to effectively manage repairs:

- First, isolate the leaking component from the rest of the system to prevent further loss of refrigerant.

- Use appropriate sealants designed for HVAC systems if the leak is minor. Follow the product's instructions carefully.

- For larger leaks or damaged components, such as corroded pipes or faulty valves, replace them entirely. Make sure any replacements meet the manufacturer's specifications.

- After repairs, perform a thorough test to ensure the leak has been completely fixed. This might involve reapplying the dye test or using the electronic detector again.

It's equally important to recharge the refrigerant correctly. Once repairs are complete, you'll likely need to recharge the system with refrigerant. Recharging must be done following manufacturer specifications to ensure correct pressures and optimal system performance. Using the wrong type or quantity of refrigerant can lead to inefficiencies and potentially damage the system.

Here's what you can do for proper recharging:

- Refer to your HVAC system's manual for the exact type and amount of refrigerant required. Each system is different, and using incorrect details can be detrimental.

- Connect the refrigerant cylinder to your system using the correct hoses and gauges. Make sure all connections are secure to avoid leaks during the process.

- Open the valves gradually to allow the refrigerant to flow into the system. Monitor the pressure closely using appropriate gauges to match the manufacturer's recommended levels.

- Once the desired pressure is achieved, close the valves and disconnect the hoses carefully. Check the system for any signs of leakage once more before considering the job complete.

Making sure you choose the correct recharge will help your HVAC system stay efficient and effective, so it can keep running smoothly for longer without any unnecessary interruptions.

Troubleshooting Electrical Components

Inspecting Electrical Connections

One of the first steps in troubleshooting electrical components in an HVAC system is inspecting the electrical connections. Loose wires, corrosion, or signs of overheating can lead to unexpected failures and even pose safety risks.

By diligently inspecting these connections, you can ensure that your HVAC unit operates smoothly and lasts for a long time. Here's what you can do:

- Start by disconnecting power from the HVAC system to safely inspect the electrical elements.

- Look for loose wires or corroded connectors, which often present as discolored or degraded material around wire attachments.

- Check for signs of overheating, such as melted insulation or burn marks on wires.

- Tighten any loose connections and clean off corrosion using a small wire brush.

- Apply a dielectric grease to prevent future corrosion and reconnect the power source after verifying all connections are secure.

Regularly performing these checks can preempt larger issues, ensuring continuity in system performance and safety.

Testing Capacitors, Relays, and Contactors

Identifying faulty electrical components like capacitors, relays, and contactors hinges heavily on understanding how to test them accurately. Using tools like multimeters or capacitance testers, homeowners and technicians can pinpoint malfunctioning parts that may disrupt the HVAC system's efficiency.

To verify proper functioning:

- **Capacitors:** Disconnect the capacitor from the circuit and use a capacitance meter to check its charge retention. A healthy capacitor should match the rated capacitance value within a 10% tolerance.

- **Relays:** Set your multimeter to continuity mode and test the relay by checking the connectivity between terminals. If there's no continuity when the relay is energized, then it's defective.

- **Contactors:** Use the multimeter to measure the voltage drop across the contactor when the system is operating. A significantly high voltage drop indicates worn contactor points.

Regular testing will identify failing components early, reducing downtime and ensuring consistent performance.

Understanding Wiring Diagrams

Navigating through wiring diagrams and electrical schematics is critical for accurate diagnosis and repair of HVAC systems. These diagrams serve as blueprints for the electrical pathways and help trace circuits effectively, vital for both diagnosing faults and implementing repairs.

Here's how to start:

- Familiarize yourself with common symbols and conventions used in HVAC wiring diagrams.

- Begin tracing from the main power supply following through to various components like compressors, fans, and thermostats.

- Identify each connection and terminal, noting where power should be present versus where it's absent.

- Cross-reference the physical layout of the system against the schematic to ensure accuracy in tracing circuits.

Understanding these diagrams empowers you to tackle complex electrical issues methodically, enhancing your ability to make precise repairs.

Following Safety Protocols

Safety cannot be overstated when working with electricity. Observing proper safety protocols minimizes risks and ensures that both the technician and the system remain protected throughout the troubleshooting process.

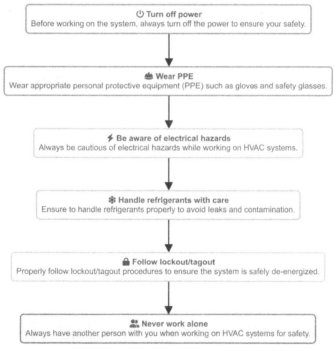

HVAC Safety Tips: Protect Yourself and Your System

Essential safety measures include:

- Always disconnect the power source before beginning any inspection or repair work.

- Use insulated tools specifically designed for electrical work.

- Wear proper Personal Protective Equipment (PPE), including gloves, goggles, and non-conductive footwear.

- Double-check that capacitors are discharged before handling them, as they can retain a dangerous charge even after the system is powered down.

- Be mindful of moisture and conductive materials that could pose electrocution and short-circuiting hazards.

From regularly inspecting electrical connections for potential issues, accurately testing key components like capacitors and relays, navigating wiring diagrams for precise repairs, to following stringent safety protocols, proficiency in these areas equips you to address and mitigate common HVAC problems.

When you have the right tools and knowledge, HVAC troubleshooting becomes much more manageable. This leads to better outcomes and gives you peace of mind. As you continue to explore and understand your HVAC system, remember that each proactive step taken today safeguards against potential challenges tomorrow.

References

U.S. Department of Energy. (2023). DOE Recognizes Eight Organizations for Excellence in Expanding Use of Smart Diagnostic Tools for Efficient HVAC Performance in Residential Buildings. Energy.gov. Retrieved from https://www.energy.gov/eere/buildings/articles/doe-recognizes-eight-organizations-excellence-expanding-use-smart

Florida Academy. (2022). 5 Common HVAC Airflow Problems and How to Fix Them?. Florida Academy. https://florida-academy.edu/5-common-hvac-airflow-problems/

Bruno, P. (2011). The importance of diagnostic test parameters in the interpretation of clinical test findings: The Prone Hip Extension Test as an example. The Journal of the Canadian Chiropractic Association, 55(2), 69. https://doi.org/10.1080/jcca.2011.69

Environmental Protection Agency. (2015). Stationary Refrigeration Leak Repair Requirements. Retrieved from https://www.epa.gov/section608/stationary-refrigeration-leak-repair-requirements

Johnson, L. (2022). Basic HVAC troubleshooting tips every homeowner should know. My Green Montgomery. Retrieved from https://mygreenmontgomery.org/2022/basic-hvac-troubleshooting-tips-every-homeowner-should-know/

U.S. Environmental Protection Agency. (2014). Heating, Ventilation and Air-Conditioning Systems, Part of Indoor Air Quality Design Tools for Schools. Retrieved from https://www.epa.gov/iaq-schools/heating-ventilation-and-air-conditioning-systems-part-indoor-air-quality-design-tools

Florida Academy. (2021). DIY HVAC Repair Guide: How to Fix Your HVAC System. Florida-Academy. https://florida-academy.edu/diy-hvac-repair-guide/

Summit College. (2021). Understanding Basic Electrical Wiring and Components in HVAC Systems. Summit College. Retrieved from https://summitcollege.edu/basic-electrical-wiring-components-hvac-systems/

U.S. Environmental Protection Agency. (2015). Stationary Refrigeration Service Practice Requirements. Other Policies and Guidance. https://www.epa.gov/section608/stationary-refrigeration-service-practice-requirements

6

HVAC System Controls

HVAC system controls, especially thermostats and automatic technologies, make sure that the temperature inside is just right while also saving energy. They do this by being the brains of heating, ventilation, and air conditioning systems.

Despite the convenience they offer, the mechanisms behind these controls can often seem elusive. Many homeowners find themselves baffled by unexplained spikes in energy bills or inconsistent indoor temperatures.

A thermostat positioned near a drafty window might cause the HVAC system to overwork, leading to higher costs and undue wear and tear. In addition, if programming is not done correctly, even the most advanced systems can end up being inefficient, resulting in unnecessary energy consumption and financial losses.

It's a good idea to have a solid grasp on both the functionality and the proper use of these systems to navigate through these challenges successfully.

The Role of Thermostats in HVAC Systems

A thermostat serves as the command center of your HVAC system, directly influencing how comfortable you feel in your home while optimizing energy efficiency.

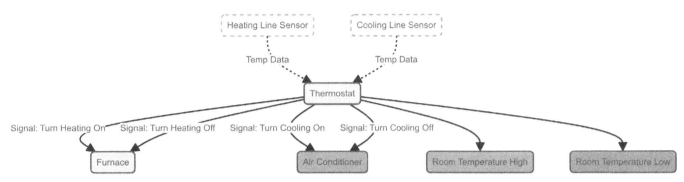

The Thermostat as the Control Center of Your HVAC System

Think of it as the conductor of an orchestra, guiding each instrument to deliver a harmony of warmth or coolness exactly when needed. By regulating the temperature in a space, thermostats ensure that heating or cooling systems activate only when necessary, helping maintain a balanced indoor climate without excessive energy consumption.

Programming thermostat schedules is where significant energy savings can be found. Most programmable thermostats let you set different temperatures for various times of the day or even days of the week. This feature can be incredibly beneficial, especially if your household follows a consistent routine.

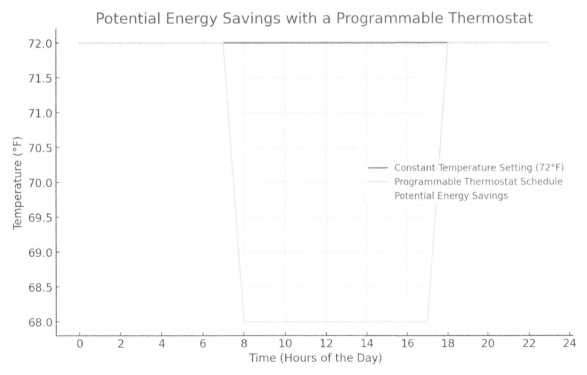

Energy Savings with a Programmable Thermostat

For instance, you can program the thermostat to reduce heating or cooling when you're away at work and then have it kick back in shortly before you return home. This way, you don't spend money maintaining an empty house at your preferred comfort level.

To make use of programmable thermostats effectively:

- Analyze your daily and weekly routines.
- Identify periods when the home is typically unoccupied.
- Adjust the thermostat settings to lower energy usage during these times.
- Experiment with various settings to find what works best for your family's schedule and comfort needs.

By thoughtfully programming your thermostat, you could save around 10% per year on heating and cooling bills simply by adjusting the temperature 7-10° F from its normal setting for eighthours a day (Energy.gov, n.d.).

When we talk about smart features in modern thermostats, we're stepping into a realm of even greater convenience and efficiency. Smart thermostats learn your patterns over time and adjust automatically.

These devices often come with remote access capabilities, allowing you to control your home's temperature via smartphone apps, regardless of where you are. You can warm up your house before getting out of bed or cool it down while on your way back from work on a blistering summer day.

Remote Control and Monitoring with a Smart Thermostat App

Additionally, many smart thermostats notify you when it's time to replace filters, improving air quality and ensuring your HVAC system runs efficiently. Some of the benefits include:

- Automated adjustments based on your habits and occupancy patterns.
- Remote access through mobile apps, giving you control from anywhere.
- Predictive maintenance alerts to keep your system in top shape.
- Integration with other smart home devices for a seamless living experience.

These features simplify managing your home environment, saving both energy and money while enhancing your overall comfort (ohio business college, 2022).

It's important to consider the placement of your thermostat for optimal HVAC performance, even though it's often overlooked. Ensuring proper placement is key to getting accurate temperature readings and avoiding unnecessary cycling of your HVAC system. Here are some key points to consider:

- Install the thermostat on an interior wall, away from direct sunlight, drafts, doorways, skylights, and windows to avoid "ghost readings."
- Make sure it's located where natural room air currents occur—warm air rising and cool

air sinking.

- Avoid placing furniture in front of or below the thermostat to ensure unobstructed airflow.

- Ensure it is easily accessible for convenient programming and adjustments.

Placing the HVAC system incorrectly can result in inefficient operation, higher energy expenses, and inconsistent indoor temperatures. By following these guidelines, your thermostat will provide more reliable temperature control, contributing to both energy savings and comfort.

Automation and Smart Control Options for HVAC Systems

Automated HVAC systems can be programmed to follow a set schedule, ensuring that your living space is always at the ideal temperature when you need it most. That means no more manually adjusting the thermostat when you leave home or come back after a long day.

Whether it's time to turn down the heat at bedtime or cool things off during the hottest part of the day, automation handles these tasks seamlessly. Here is what you can do to achieve the goal:

- Identify the key times during the day when you are typically home and away. Program your HVAC system to adjust settings accordingly.

- Use the app or interface provided by your HVAC system to set up these schedules. Most modern systems offer intuitive controls that make this process simple.

- Take advantage of remote monitoring features. If your plans change, you can adjust your HVAC settings from anywhere using your smartphone.

- Enable adaptive adjustments. Many automated systems learn your preferences over time, optimizing their operation based on your daily routines and the changing seasons.

Smart control features enable integration with other smart home devices, enhancing overall comfort and convenience. These integrations not only make life more comfortable but also maximize energy efficiency.

HVAC systems that are automated have the ability to adapt to different environmental conditions, ensuring that indoor comfort is maintained in an efficient manner.

Let's say there's an unexpected spike in outdoor temperatures. A smart HVAC system can detect this and adjust its output to maintain a consistent indoor climate without you lifting a finger. This adaptability ensures that your home remains comfortable regardless of sudden weather changes.

Leveraging automation reduces human error and ensures consistent and reliable operation of HVAC equipment. How many times have we forgotten to turn off the air conditioning before leaving for vacation, only to return to staggering energy bills? Automation eliminates these oversights by maintaining preset parameters, ensuring that your HVAC system runs optimally without wasteful practices.

The beauty of automation lies in its ability to integrate with various smart home platforms, providing users with a cohesive and unified experience.

For example, when your HVAC system detects that you're away from home, it can automatically switch to an energy-saving mode. It will also keep track of your return using the location services on your smartphone. Just imagine, these smart thermostats are so advanced that they can actually analyze your preferences and patterns. They use machine learning algorithms to fine-tune their operation to perfectly match your lifestyle. It's like having a thermostat that understands you like a person would.

Moreover, automated systems often come with diagnostic tools that alert you to potential issues, prompting timely maintenance and avoiding costly repairs. These alerts can be sent directly to your phone or email, providing peace of mind and ensuring the longevity of your system.

Thanks to advanced technology, managing complex HVAC systems has become much easier. In fact, according to recent studies on smart home automation systems, introducing advanced control features into residential HVAC systems can lead to substantial energy savings and improved system performance (Bachler et al., n.d.). Automation not only provides personal comfort, but also supports the larger objective of sustainability.

What's really cool about these intelligent systems is that they're designed to work with minimal intervention. Once the initial setup is complete, you can largely step back and allow the system to handle the nuances of climate control. This is particularly beneficial if you're juggling hectic schedules or simply looking to minimize household chores.

In addition, when your HVAC system integrates with other smart devices, it becomes a seamless part of a connected ecosystem.During winter, smart blinds can open to let in natural sunlight, reducing the heating load on your system. Similarly, smart vents can adjust airflow based on occupancy detected by motion sensors, ensuring that energy isn't wasted in unoccupied spaces.

An often overlooked yet crucial benefit of smart controls is their potential to enhance indoor air quality. Modern systems come equipped with sensors that can detect pollutants and allergens, adjusting ventilation rates accordingly. This ensures that the air you breathe indoors remains fresh and healthy, which is especially important for households with members suffering from respiratory issues.

It's worth noting that leveraging automation doesn't mean relinquishing control entirely. Users still retain the option to override automated settings through manual inputs when necessary. This flexibility ensures that the system caters to personalized comfort while adhering to efficient operation standards.

Energy Management Systems and HVAC Performance Optimization

Energy management systems play an indispensable role in optimizing HVAC performance. These systems function as the brain of your heating, ventilation, and air conditioning apparatus, ensuring that it operates as efficiently as possible while maintaining the desired comfort levels within buildings.

When homeowners and facility operators use these advanced management tools, they can not only save money on their energy bills, but also make a positive impact on the environment.

Energy management systems facilitate monitoring energy usage, identifying inefficiencies, and implementing energy-saving strategies. Think of this system as a meticulous auditor, or a vigilant friend constantly reminding you to turn off lights when you leave a room or adjust the thermostat when you step out for the day.

Importantly, here is what you can do to achieve optimal usage:

- Regularly check your energy consumption patterns using the system's interface.

- Identify any sudden spikes in energy use which might indicate inefficient equipment or misuse.

- Implement suggestions from the system such as adjusting the thermostat, sealing duct leaks, or scheduling regular maintenance checks for your HVAC units.

Energy management systems offer valuable data that can provide insights into HVAC performance trends and patterns, helping with informed decision-making. The beauty of data lies in its ability to tell us stories about our behaviors and practices that we might not notice otherwise.

For instance, you might start noticing that certain times of day see a significant increase in energy use. Or perhaps some parts of your building or home are consistently consuming more energy than others.

Equipped with this knowledge, you can make better decisions about equipment upgrades, operational changes, or even behavioral adjustments. Such informed decisions lead to a more efficient operation and help avoid the unnecessary strain on both your pockets and the planet.

Another crucial aspect of these systems is that they allow for adjusting settings based on energy usage patterns, leading to cost savings and reduced environmental impact. If your system reveals that your HVAC is overworking during times when no one is at home or in the building, you can set it to switch to an energy-saving mode during those periods.

Adjusting your HVAC operations doesn't just save money; it also means you're actively participating in reducing greenhouse gas emissions. Having preset schedules tailored to your actual usage patterns ensures that energy isn't wasted, making your entire HVAC operation smarter and greener (Energy.gov, n.d.).

Adding energy management tools to HVAC controls also makes the whole system more efficient and environmentally friendly. This integration doesn't just mean adding a new piece of hardware or software; it changes the way your HVAC system interacts with and reacts to its surroundings in a fundamental way.

Smart thermostats and automated dampers, for instance, can dynamically adjust operations based on real-time data, weather conditions, and occupancy levels. This means that your system is always working optimally, neither overburdened nor underperforming.

Additionally, controlling your HVAC system through an integrated energy management framework can detect and prevent issues before they become major problems.

Imagine a scenario where a part of your system is starting to falter. Traditional setups might miss the early signs, but an integrated system alerts you instantly, allowing for timely intervention. Regular maintenance prompts, system diagnostics, and remote management capabilities mean your HVAC is always performing at its best, all with minimal manual intervention required (Murphy, 2023).

Understanding the role of controls in HVAC systems and their impact on performance has been the focal point of this chapter. However, you might be concerned about the initial investment and complexity of setting up these advanced systems.

While there are upfront costs and a learning curve, the long-term benefits—manifested as lower utility bills, improved comfort, and greater environmental responsibility—make it a worthwhile endeavor.

On a broader scale, embracing these innovations contributes to a collective effort towards energy efficiency and sustainability. As our homes and buildings become smarter and more responsive, we pave the way for a future where managing our living environments is intuitive and responsible.

Integrating HVAC Controls with Building Management Systems

Integrating HVAC controls with building management systems (BMS) is an essential step in modern facility management. This integration enables centralized control and monitoring of HVAC operations in larger facilities, making management tasks more efficient and streamlined.

Think of it as having a central command center for your entire building's climate systems. This means that instead of managing each HVAC unit individually, everything can be controlled from one place, providing unparalleled oversight and convenience.

With centralized control, facility managers can monitor the status of all HVAC units at once, ensuring they function optimally and troubleshooting issues before they escalate into major problems. This centralized approach not only simplifies daily operations but also provides a strategic advantage in maintaining system performance and extending equipment lifespan through timely interventions (jenks2026, 2024).

Beyond HVAC alone, integrating these controls with other building systems like lighting fosters greater efficiency and occupant comfort. When HVAC systems work in tandem with lighting, it's easier to create a cohesive environment that maximizes both energy savings and user satisfaction.

For instance, during the day when natural light floods through windows, the system can automatically adjust interior lights and HVAC settings to save energy without compromising comfort.

The integration of various building systems allows for adaptive responses based on real-time data.

Picture yourself walking into a conference room and instantly, the lights flicker on and the temperature adjusts to your liking, as if the room is responding to your presence. This seamless coordination leads to significant energy reductions and an overall improved indoor experience.

Field research has shown such strategies can reduce HVAC energy use by up to 47%, demonstrating the substantial impact of integrated controls (Slipstream, n.d.).

Another significant advantage of integrating HVAC controls with BMS is enhanced data sharing between systems. This connectivity opens possibilities for advanced analytics and proactive maintenance.

Data collected from HVAC operations, when analyzed correctly, can reveal patterns and predict potential failures before they happen. Proactive maintenance strategies, driven by this data, ensure that minor issues are addressed before turning into costly repairs. This predictive maintenance approach not only saves money but also reduces downtime, keeping building operations smooth and uninterrupted (Stromquist, 2023).

Real-life examples show how data-driven insights can transform HVAC system management.

If sensors indicate that a particular HVAC unit is consuming more energy than usual, the system can flag this anomaly, triggering a preemptive inspection. Facility managers can then address the problem early, avoiding expensive emergency repairs. Furthermore, historical data trends help optimize system settings over time, ensuring peak performance under varying conditions.

Integration also facilitates optimized operation through predictive analytics. These systems can leverage real-time data to make informed decisions about how to run building systems most efficiently.

If weather forecasts predict a heatwave, the system can pre-cool the building during off-peak hours to reduce stress on the HVAC system during peak usage times. Such predictive adjustments maintain comfort while minimizing energy consumption peaks, which can be costly.

There are many other applications where predictive capabilities can be used, not limited to weather-related optimizations. Analyzing data continuously allows for optimizing system operations to achieve optimal efficiency.

Over time, the system "learns" the building's unique characteristics and occupancy patterns, enabling it to make smarter adjustments autonomously. This ongoing optimization ensures that energy use is constantly being refined, promoting sustainability and cost-efficiency.

Incorporating these advanced systems may seem complex, but the benefits far outweigh the initial investments. With reduced operational costs, enhanced occupant comfort, and improved energy efficiency, integrating HVAC controls with BMS represents a forward-thinking approach to building management.

References

COVID-19 and its social implications

Ohio Business College. (2022). 4 Reasons to Use a Smart Thermostat in Your Home. Ohio Business College. Retrieved from https://obc.edu/4-reasons-to-use-a-smart-thermostat-in-your-home/

Asim, N., Badiei, M., Mohammad, M., Razali, H., Rajabi, A., Haw, L. C., & Ghazali, M. J. (2022). Sustainability of Heating, Ventilation and Air-Conditioning (HVAC) Systems in Buildings—An Overview. International Journal of Environmental Research and Public Health, 19(2), 16. https://doi.org/10.3390/ijerph19021016

Jenks, J. (2024). Building Management Systems (BMS): Revolutionizing Modern Building Management. Green.org. Retrieved from https://green.org/2024/01/30/building-management-systems-bms/

Stromquist, E. (2023). Unlocking the Benefits of Building Automation Control Systems. ControlTrends. Retrieved from https://controltrends.org/hvac-smart-building-controls/building-automation-and-integration/10/unlocking-the-benefits-of-building-automation-control-systems/

Borza, D., & Popa, A. (2023). An upgraded approach in sustainable energy. Applied Energy, 310, 121921. https://doi.org/10.1016/j.apenergy.2023.121921

Mississippi State University Extension Service. (n.d.). Energy Efficient Homes: Programmable Thermostats. Retrieved from http://extension.msstate.edu/publications/energy-efficient-homes-programmable-thermostats

Skeledzija, N., Cesic, J., Koco, E., Bachler, V., Vucemilo, H. N., & Džapo, H. (N.D.). Smart home automation system for energy efficient housing. IEEE Conference Publication. https://ieeexplore.ieee.org/document/6859554/

Journal of Youth and Adolescence, 49(5), 1207-1225. https://doi.org/10.1007/s10964-020-01265-0

Murphy, D. (2023). Optimizing energy efficiency with HVAC. ITI Technical College. https://iticollege.edu/blog/the-role-of-hvac-technicians-in-sustainable-building-design/

Energy.gov. (n.d.). Programmable Thermostats. Retrieved from https://www.energy.gov/energysaver/programmable-thermostats

U.S. Department of Energy. (n.d.). Energy Management Systems: Maximizing Energy Savings (Text Version). Energy.gov. https://www.energy.gov/scep/energy-management-systems-maximizing-energy-savings-text-version

Slipstream. (n.d.). Integrating Lighting and HVAC Controls: Solutions for High Performance Buildings. Retrieved from https://slipstreaminc.org/research/integrating-lighting-and-hvac-controls-solutions-high-performance-buildings

7
Energy Efficiency in HVAC

Despite the essential role HVAC systems play, they often come with drawbacks, notably high energy use and substantial utility bills. Traditional HVAC units can be highly inefficient, wasting energy and money due to outdated technology and poor maintenance practices.

An old air conditioning unit might struggle to keep up with demand on a hot summer day, consuming excessive amounts of electricity and driving up monthly utility expenses. Likewise, a neglected furnace can run inefficiently in the winter, leading to higher heating costs.

These challenges highlight the need for more efficient solutions that can alleviate the financial burden on homeowners and contribute positively to environmental sustainability.

The quest for energy efficiency in HVAC systems has become increasingly important as homeowners look for ways to reduce energy consumption and costs while maintaining optimal indoor comfort. How can we achieve this balance by making HVAC systems smarter and more efficient?

High-Efficiency HVAC Equipment Options

High-efficiency HVAC units are changing the game in home heating and cooling. With advanced technology at their core, these units are designed to minimize energy waste significantly. By doing so, they not only help lower utility costs for homeowners but also contribute to broader environmental goals.

Picture a cozy winter evening where your furnace operates at peak efficiency, keeping you snug and toasty while conserving energy. These systems work by using smart thermostats, variable-speed motors, and enhanced heat exchangers to maximize performance according to current needs.

Now, let's take a closer look at the financial side of things. While high-efficiency HVAC equipment might come with a higher upfront cost, consider it an investment rather than an expense. Over time, the savings on your energy bills can add up substantially.

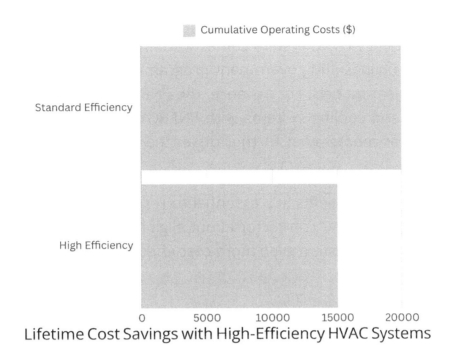
Lifetime Cost Savings with High-Efficiency HVAC Systems

According to the U.S. Department of Energy's eCompass guide, switching to high-efficiency equipment can reduce electricity use by 50% for electric heating and cooling systems (U.S. Department of Energy and ENERGY STAR®, n.d.).

This means that over several years, you're likely to recoup your initial investment and then some due to lower utility bills and fewer repair costs.

Moreover, these systems usually have a longer lifespan, translating to increased system durability and reliability over the long haul.

However, getting the most out of these high-efficiency units isn't just about installing them and forgetting about them. Proper maintenance is crucial to ensure optimal performance and energy savings over time. Here's what you can do to keep your system running smoothly:

- Change your air filters every one to three months, depending on usage.
- Schedule annual tune-ups with a certified HVAC technician to inspect and clean critical components.
- Keep the area around your outdoor unit clear of debris, leaves, and shrubs.
- Regularly check and clean the condensate drain to prevent clogging and mold growth.

By following these simple maintenance steps, you can ensure that your high-efficiency HVAC system operates at peak performance, providing both comfort and savings year after year.

Investing in a high-efficiency HVAC system comes with another perk: potential rebates and incentives. Many local utilities and government programs offer financial incentives for upgrading to energy-efficient models. For instance, the ENERGY STAR® program reports that replacing old heating and cooling systems with ENERGY STAR®-certified equipment can qualify homeowners for rebates that further offset the initial investment costs (ENERGY STAR®, n.d.).

To take advantage of these benefits, it's essential to research available programs in your area and consult with your HVAC contractor about eligible models. Doing so can make the transition to a high-efficiency system even more cost-effective and rewarding.

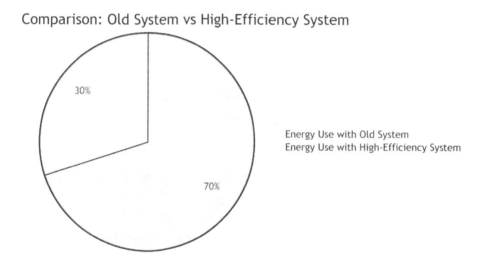

Energy Savings Potential with High-Efficiency HVAC Systems

Energy Recovery Ventilation Techniques

Energy recovery ventilators (ERVs) harness the energy from conditioned air inside a building and reuse it to pre-condition incoming air, effectively reducing the workload on your HVAC system. This technique ensures that you're not simply throwing away all the energy used to heat or cool indoor air as it exits your home (AHRI). Instead, the ERV captures that energy and uses it to warm or cool the incoming fresh air, making the entire HVAC system more efficient. It's akin to recycling energy, ensuring nothing goes to waste.

Understanding this process can significantly impact how you approach HVAC maintenance and upgrades. By implementing ERVs, you achieve better control over your home's temperature with less effort from your heating and cooling systems. This not only prolongs the life of your HVAC unit but also translates into noticeable savings on energy bills.

Moreover, while energy efficiency is critical, maintaining good indoor air quality shouldn't be compromised. ERVs play a pivotal role here too.

Unlike traditional systems that might merely circulate indoor air, ERVs ensure a continuous flow of fresh air into your living spaces without letting go of already-used energy (Building Science Education). Fresh air is vital for diluting pollutants such as moisture, dust, pollen, and other particulates that accumulate indoors.

By consistently bringing in fresh outdoor air and saving energy at the same time, you can reduce the health risks linked to poor indoor air quality, such as asthma and allergies.

Understanding the intricacies of ERVs is essential to fully harness their benefits. They come equipped with heat exchangers and fans that facilitate the transfer of heat (and in some cases, moisture) between outgoing and incoming air streams.

This continuous airflow exchange ensures pollutants are expelled while the comfortable, conditioned environment inside remains largely unaffected.

Essentially, it's a win-win scenario—enhancing indoor air quality while keeping energy consumption low.

Whether it's blazing hot summers or freezing winters, ERVs help maintain a stable and comfortable indoor climate more efficiently. This is because the technology capitalizes on the contrasting temperatures of outgoing and incoming air streams to ease the burden on your HVAC system (Institute for Market Transformation (IMT), 2023).

In summer, the cooler indoor air helps pre-cool the incoming hot and humid air. Conversely, in winter, the warm indoor air helps to pre-warm the incoming cold air. The larger the temperature difference between these airflows, the more effective the energy recovery process.

As we're discussing the broader implications and advantages, it's crucial to highlight the substantial contributions of ERVs toward overall energy savings and environmental sustainability. Modern HVAC systems, when supplemented with ERV technology, perform at optimal levels due to the reduced need for constant heating or cooling.

This streamlined performance leads to significant cuts in energy consumption and operating costs over time (HVAC Energy Recovery Ventilation | Building Science Education).

Here's what you can do to optimize HVAC system performance using ERV techniques:

- Ensure your ERV system is correctly sized for your home or building. This involves calculating the volume of ventilation required and selecting an ERV model that can handle this load efficiently.

- Install the ERV in a location where the fresh air intake and exhaust ducts are relatively close to each other. This reduces ductwork complexity and enhances system efficiency.

- Maintain regular inspections and cleaning schedules for your ERV unit to ensure it operates at peak performance. Removing dust and particulate build-up from filters and heat exchangers is crucial.

- In climates with significant seasonal variations, consider models that can switch functionalities between heating and cooling modes efficiently. Some advanced models come with built-in controls for adaptive defrosting and economizing features, which are helpful in managing extreme conditions (Institute for Market Transformation (IMT), 2023).

Studies show that homes with ERVs not only save on utility bills but also contribute to lower greenhouse gas emissions (Institute for Market Transformation (IMT), 2023). These findings highlight the dual benefits of personal cost savings and positive environmental impact, bridging the gap between individual concerns and societal responsibilities.

Significance of Seasonal Energy Efficiency Ratio (SEER)

Let's move on to the Seasonal Energy Efficiency Ratio, or SEER. This rating is a simple yet powerful indicator that helps us gauge how efficiently an air conditioner or heat pump can operate.

The higher the SEER rating, the more efficient the system. Think of it like miles per gallon in your car—the more miles you can get out of each gallon, the less fuel you need over time. Similarly, a higher SEER rating means your HVAC system needs less energy to provide the same level of cooling.

A system with a higher SEER rating not only reduces energy consumption but also lowers operating costs throughout its lifespan. If you have an HVAC system with a SEER rating of 10 and one with a rating of 16, the latter will offer significant energy savings over years of use, leading to lower utility bills.

In fact, modern units with high SEER ratings are sometimes 20-40% more efficient than older models (Breidinger, 2023). These savings accumulate, making a higher initial investment worthwhile for many homeowners.

However, simply choosing an HVAC system with a high SEER rating isn't enough to guarantee peak performance. Regular maintenance and proper installation practices are indispensable. Here's what you can do to ensure your HVAC unit operates at its highest efficiency:

- Schedule annual maintenance checks before the peak heating or cooling season begins. A professional can examine your unit for any wear and tear and make necessary adjustments.

- Replace air filters every 1-3 months during periods of heavy use. A clogged filter restricts airflow, forcing the system to work harder.

- Ensure that outdoor units are free from debris and have enough clearance around them to allow for proper airflow.

- Verify that ductwork is sealed and insulated. Leaky ducts can result in significant energy losses, negating the benefits of a high SEER rating.

Understanding SEER values can be a game-changer when you're considering HVAC upgrades. Higher SEER ratings translate to better efficiency and lower operational costs. But these improvements come at a price, so balancing budget constraints with sustainability goals is essential.

When shopping for a new system, consider the long-term savings a high-SEER unit offers against its initial cost. Many units within the 14 to 21 SEER range can be optimal choices for residential use, with some newer models reaching even higher efficiencies (Murphy, 2016).

When evaluating potential HVAC upgrades, it's useful to compare units based on their SEER ratings. Many units now come with a yellow and black EnergyGuide label that clearly displays this information. If not, a quick reference to the system's model number usually includes the SEER rating (Wright-Hennepin Cooperative Electric Association, 2013).

To get the most accurate comparison, look at the SEER alongside other factors like estimated annual energy costs, maintenance requirements, and warranty terms.

The importance of SEER becomes clear when you consider the potential environmental impact. Higher SEER ratings mean less energy consumption, which typically translates to reduced greenhouse gas emissions. With climate change concerns becoming more urgent, opting for a high-efficiency HVAC system is a responsible choice that aligns personal comfort with global sustainability. Lower energy usage reduces both household expenses and ecological footprints, making it a win-win situation for everyone involved.

Integrating Renewable Energy Sources in HVAC Designs

Renewable energy sources, such as solar panels and geothermal heat pumps, can significantly supplement traditional HVAC systems. By harnessing the power of the sun or the constant temperature of the earth, these technologies reduce our reliance on fossil fuels while also lowering carbon emissions—an essential step towards mitigating climate change.

For instance, solar panels can be installed on rooftops to capture sunlight and convert it into electricity, which then powers the HVAC system. Similarly, geothermal heat pumps can use the stable underground temperatures to provide heating in the winter and cooling in the summer, doing so with greater efficiency than conventional systems (Asim et al., 2022).

Here's a straightforward guide to incorporating these renewable energy sources effectively:

- Assess your property to determine suitability for solar panels or geothermal systems. This involves evaluating roof space for solar installation or conducting soil testing for geothermal feasibility.

- Consult with professionals to design a system that meets your specific needs. They can help you choose the right size and type of panels or pumps.

- Consider any local incentives or rebates available for renewable energy installations. Financial assistance can significantly offset initial costs.

- Plan for the integration of the renewable energy system with your existing HVAC setup. This might require upgrading your current systems or installing additional components for seamless operation.

- Ensure proper maintenance by scheduling regular check-ups. Solar panels need cleaning to maintain efficiency, and geothermal systems should be inspected periodically for optimal performance.

When renewable energy technologies are integrated with HVAC systems, the benefits extend far beyond just the positive impact on the environment. Homeowners can enjoy significant cost savings over time due to reduced energy bills. Depending on the size and capacity of the system, solar energy can cover a substantial portion, if not all, of your HVAC needs. This leads to energy independence, making you less susceptible to energy price fluctuations and supply issues.

Additionally, employing renewable sources showcases a commitment to environmental responsibility, setting an example for others in your community.

You should also consider proper sizing and design when incorporating renewable energy sources into HVAC systems to maximize efficiency and performance (Mahmood et al., 2023).

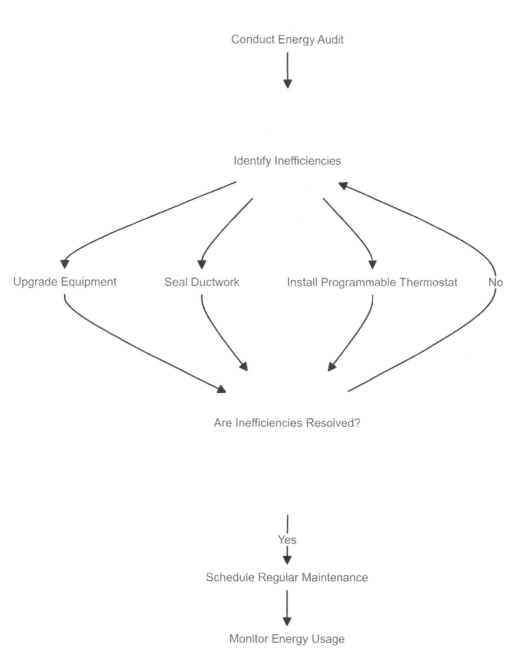

Steps to Optimize HVAC Energy Efficiency

An oversized system might lead to unnecessary expenses and inefficiencies, while an undersized one won't meet your building's demands satisfactorily. Here's what you can do to ensure optimal sizing and design:

- Start by calculating your home's energy consumption patterns to understand peak usage times and overall requirements.

- Use professional software or consult an expert to model different scenarios and predict how much energy the proposed system will generate versus the demand.

- Select equipment that fits within these calculated parameters. For solar panels, this means choosing the right number of panels and strategic placement to capture maximum sunlight. For geothermal systems, this could mean adjusting the depth and length of underground loops.

- Plan for future scalability. Your energy needs might grow, so it's wise to consider a system that can be easily expanded without major overhauls.

- Optimize the design for environmental factors like shading and seasonal variations. Solar panel angles and directions must be adjusted to match the changing position of the sun throughout the year.

Finally, combining renewable energy with other energy-efficient HVAC solutions forms a holistic approach to sustainability. This strategy enhances your home's green credentials and contributes to a greener and more resilient built environment. By implementing a mix of renewable energy technologies and high-efficiency HVAC systems, you create a synergy that leverages the strengths of each component.

For instance, pairing solar panels with energy-efficient air conditioning units ensures that the clean energy generated is used effectively, minimizing wastage and maximizing performance. It also helps in creating a built environment that is better equipped to handle future energy challenges.

Remember, starting small can lead to significant changes. Each step taken towards integrating renewable energy and improving HVAC efficiency brings us closer to a more sustainable and responsible way of living. So, assess your options, seek professional advice, and embrace the future of energy-efficient HVAC systems today.

References

WBDG. (n.d.). High-performance HVAC. Retrieved from https://www.wbdg.org/resources/high-performance-hvac

NEIF. (2023). Cooling Season 101: Understanding SEER This Summer. Retrieved from https://www.neifund.org/cooling-101-understanding-seer-ratings/

U.S. Department of Energy &ENERGY STAR®. (Each year in the U.S., three million heating and cooling systems are replaced and $14 billion is spent on HVAC services or repairs). Retrieved from https://rpsc.energy.gov/tech-solutions/hvac

Institute for Market Transformation (IMT). (2023). Very High Efficiency Commercial HVAC System Design Specification and Guidelines. IMT. https://imt.org/business-practices/very-high-efficiency-hvac/spec/

Ahrinet. (n.d.). Air-to-air energy recovery ventilators. Retrieved from https://www.ahrinet.org/scholarships-education/education/contractors-and-specifiers/hvacr-equipmentcomponents/air-air-energy-recovery-ventilators-ervs

Ali, B. M., & Akkaş, M. (2023). The Green Cooling Factor: Eco-Innovative Heating, Ventilation, and Air Conditioning Solutions in Building Design. Applied Sciences, 14(1), 195. https://doi.org/10.3390/app14010195

Asim, N., Badiei, M., Mohammad, M., Razali, H., Rajabi, A., Haw, L. C., & Ghazali, M. J. (2022). Sustainability of Heating, Ventilation and Air-Conditioning (HVAC) Systems in Buildings—An Overview. International Journal of Environmental Research and Public Health, 19(2), 1-16. https://doi.org/10.3390/ijerph19021016

Wright, M. (n.d.). Air conditioning SEER ratings explained. Willamette Valley Electric Cooperative. Retrieved from https://www.whe.org/blog/air_conditioning_seer_ratings_explained.html.

Murphy, D. (2022). Learn about the Seer rating scale from an HVAC school. ITI Technical College. Retrieved from https://iticollege.edu/blog/learn-about-the-seer-rating-scale-from-an-hvac-school/

ENERGY STAR®. (n.d.). When is it time to replace?. https://www.energystar.gov/saveathome/heating-cooling/replace

U.S. Department of Energy. (n.d.). HVAC energy recovery ventilation. Retrieved from https://bsesc.energy.gov/energy-basics/hvac-energy-recovery-ventilation

8

Air Quality in HVAC Systems

We spend most of our time indoors, whether it's at home, at work, or in other enclosed spaces, making the quality of the air we breathe inside pivotal to our overall well-being. Indoor air pollutants are more common than many people realize and can originate from various sources within our homes.

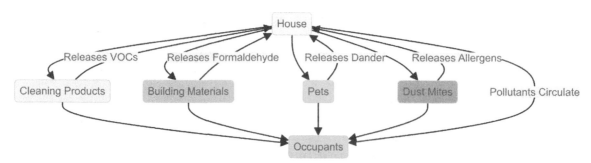

Sources and Circulation of Indoor Air Pollutants

For instance, mold thrives in damp conditions and can be found in poorly ventilated bathrooms or basements. Dust mites, which are microscopic creatures living in household dust, thrive in warm, humid environments and are known to trigger allergic reactions and asthma.

Products such as paints, varnishes, and even everyday cleaning supplies emit Volatile Organic Compounds (VOCs), which can pose a significant challenge. Inhaling these compounds over time can result in chronic respiratory problems. Poor ventilation worsens these problems by trapping pollutants indoors, resulting in higher concentrations compared to the usual levels found in outdoor air.

Understanding Indoor Air Pollutants

Indoor air quality isn't just about comfort; it's fundamentally tied to our health and well-being. Many of us spend a substantial amount of time indoors, whether we're at home relaxing or working in an office.

Because we breathe indoor air almost constantly, it's crucial to understand what might be polluting it—and how we can fix that.

Let's explore some of the most common indoor air pollutants. There's mold, dust mites, and volatile organic compounds (VOCs).

Mold thrives in damp environments and if left unchecked, can contribute to serious respiratory problems and allergies. You might find mold in areas with poor ventilation such as bathrooms and basements.

Dust mites, on the other hand, are tiny creatures that live in household dust. They can trigger asthma and allergic reactions, making them a significant concern for many.

VOCs are a bit different. These compounds are released into the air from products like paints, varnishes, and adhesives. Even everyday items like markers and cleaning products can emit VOCs.

Breathing in these chemicals over time can lead to chronic respiratory conditions and even exacerbate existing health problems (EPA et al., 2014). While not everyone is equally sensitive to these pollutants, it's essential to minimize their presence in our living spaces.

Understanding where these pollutants come from gives us a better chance of controlling them. Mold, for instance, often stems from excessive moisture or leaks. Keeping your home's humidity in check and addressing any water damage promptly can prevent mold growth.

Dust mites thrive in warm, humid environments and are commonly found in bedding, upholstered furniture, and carpets. Regular cleaning and using allergy-proof covers can mitigate their impact.

Household cleaning products contain a cocktail of chemicals that contribute significantly to indoor air pollution. Tobacco smoke is another notorious source of indoor air contaminants, filled with harmful substances that affect both smokers and non-smokers alike. Moreover, inadequate ventilation can trap pollutants inside, leading to higher indoor pollutant levels than one might find outdoors (Reducing Your Exposure to Indoor Air Pollution, n.d.).

One effective way to combat indoor air pollutants is through HVAC filters and air purifiers. These devices are designed to capture and remove contaminants from the air, thereby improving overall air quality. For homeowners looking to take charge of this aspect, here are some practical steps you can take:

- Choose high-efficiency particulate air (HEPA) filters for your HVAC systems. These filters are capable of trapping smaller particles that would otherwise pass through regular filters.

- Consider investing in air purifiers with HEPA and activated carbon filters. The combination helps in removing both particulate matter and gaseous pollutants such as VOCs.

- Keep your windows open whenever possible to allow fresh air to circulate and dilute indoor pollutants.

- If you're remodeling or redecorating, opt for low-VOC products to reduce the risk of long-term exposure.

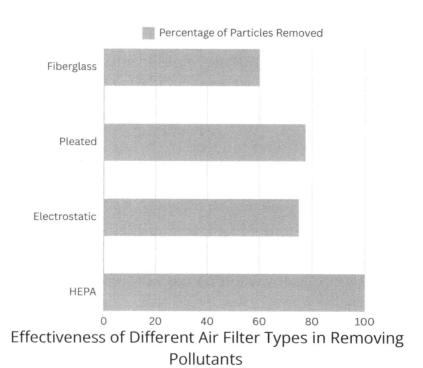

Effectiveness of Different Air Filter Types in Removing Pollutants

Regular maintenance of HVAC systems can play a key role in preventing the buildup of pollutants. Often, we overlook the importance of routine checks and cleaning, but these simple actions can significantly enhance indoor air quality over time.

Here's a guideline since it's vital to keep your HVAC system running efficiently:

- Schedule regular professional inspections to ensure that the system is functioning correctly and to identify any issues early on.

- Replace or clean your filters according to the manufacturer's recommendations. This prevents the system from circulating contaminated air.

- Ensure that the ducts are clean and free from blockages. Dirty ducts can house mold, dust, and other harmful particles.

- Keep an eye on moisture levels in your HVAC system. Condensation in the system can become a breeding ground for mold, which then gets distributed throughout your home.

Taking care of indoor air pollutants goes beyond just checking off a list; it's about making sure you and your loved ones have a healthier environment. When we think about the sources of indoor air pollution and how to mitigate them, HVAC systems stand out as a powerful tool in this fight. By actively managing and maintaining these systems, we can make a noticeable difference in the quality of the air we breathe every day.

Effective Ventilation Strategies

Improving indoor air quality through proper ventilation is crucial for a healthy living environment. It's not just about comfort; it's about ensuring that the air we breathe inside our homes remains as pollutant-free as possible. With a well-ventilated space, you're essentially diluting indoor pollutants by introducing fresh outdoor air.

The science behind this is straightforward and backed by extensive research (EPA et al., 2014). When you allow clean air to flow into your home, it mixes with the stagnant indoor air, reducing concentrations of pollutants like VOCs, carbon dioxide, and other airborne contaminants.

Ventilation strategies are diverse and can be tailored to fit different needs and setups. One of the most direct ways to improve air circulation is using exhaust fans, air vents, and air exchangers.

Exhaust fans come in handy in spaces like bathrooms and kitchens where moisture and odors tend to build up. Air vents and air exchangers work on a broader scale, helping to move old, stale air out and bring fresh air in.

To effectively use these ventilation systems and achieve optimal results in maintaining indoor air quality:

- Install exhaust fans in high-moisture areas such as bathrooms and kitchens. These fans should expel air directly to the outside rather than into the attic or other parts of the house.

- Use range hoods over your stove when cooking to capture grease, smoke, and cooking odors.

- Incorporate air exchangers that can balance the airflow between indoors and outdoors. This is particularly useful in airtight homes that don't naturally breathe well.

Balanced ventilation systems take this a step further by maintaining a consistent and controlled amount of fresh air brought in while ensuring adequate filtration. Such systems are designed to adjust the intake and exhaust rates automatically, responding to the conditions inside the home. This balance is essential because too much ventilation could lead to energy loss, while too little could compromise air quality.

Here are steps you can follow to set up and maintain a balanced ventilation system:

- Ensure your ventilation system includes both an intake and exhaust mechanism that work simultaneously to maintain equilibrium.

- Regularly check and clean the intake and exhaust ports to prevent blockages that could hinder airflow.

- Use high-efficiency particulate air (HEPA) filters within the system to trap contaminants before they enter or leave the living space.

- Schedule periodic inspections and maintenance checks to keep the system running smoothly and efficiently.

Now, let's talk about why all of this is worth your attention. Without effective ventilation, indoor air quality can deteriorate rapidly. Modern homes, built to be energy-efficient, often suffer from poor natural ventilation due to tight construction that limits the natural flow of air. This can lead to a buildup of indoor pollutants which might cause health issues ranging from minor irritations to serious respiratory conditions over time (Association, n.d.).

Moreover, better ventilation isn't just beneficial for human occupants; it's great for the home itself. Moist, stale air can lead to mold growth, which can degrade building materials, causing structural damage over time. It can also be a breeding ground for dust mites and other allergens that may trigger allergic reactions.

The bottom line is clear—effective ventilation has a profound impact on indoor air quality and, consequently, on our health and well-being.

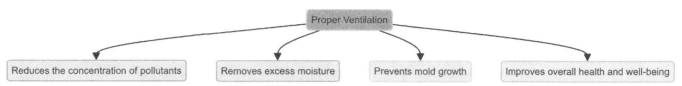

The Importance of Ventilation for Healthy Indoor Air

Advanced Air Filtration Methods

For improving indoor air quality through HVAC systems, various air filtration methods are essential in trapping and removing airborne contaminants. Let's explore some of the most effective options available.

HEPA filters stand out as one of the most highly efficient options for capturing particles as small as 0.3 microns. These filters are particularly effective in removing common household pollutants such as dust, pollen, and pet dander.

Choosing a HEPA filter is a great way for homeowners to improve the air quality in their homes. Here's a step-by-step guide to help you achieve this:

- Choose a HEPA filter that fits your HVAC system.
- Ensure it has a seal to prevent any air from bypassing the filter.
- Install the filter following the manufacturer's guidelines.
- Replace the filter as recommended, generally every six months or more frequently if it appears dirty.

By adhering to these steps, you can significantly improve the efficiency of your HVAC system and ensure it effectively captures harmful particulates (EPA, OAR, 2022).

Next up, electrostatic air filters offer another robust solution. These filters work by using an electric charge to attract and trap particles. They can enhance air quality by capturing not only dust and allergens but also smaller particles that might otherwise slip through other types of filters.

The key advantage here is that these filters are washable and reusable, making them a cost-effective choice for many homes. This method reduces waste and provides an ongoing solution for maintaining air quality. To maximize their effectiveness, ensure regular cleaning as per the manufacturer's instructions to avoid buildup that could reduce airflow.

UV germicidal lamps add another layer of defense by targeting bacteria, viruses, and mold spores circulating in the air. Unlike traditional filters, UV lamps work by emitting ultraviolet light that kills or deactivates these harmful microorganisms.

This technology is notably used in both residential and commercial settings to maintain healthier environments. For those looking to incorporate UV germicidal lamps into their HVAC systems, here are a few practical steps:

- Place the lamp near the coils of the HVAC unit where it can treat the air passing through.

- Ensure the lamp's exposure time is sufficient to be effective, typically requiring continuous operation.

- Regularly check and replace UV lamps as they lose intensity over time, usually after about 9,000 hours of use.

- Follow safety precautions during installation and maintenance to avoid direct skin or eye exposure to UV light, which can be harmful.

By integrating UV germicidal lamps, you can substantially reduce the concentration of airborne pathogens, leading to improved indoor air quality (ASHRAE, n.d.).

Ensuring these filtration methods perform effectively over time is heavily dependent on maintenance. Regularly replacing air filters and conducting routine HVAC maintenance can significantly enhance the filtration performance. Dirty or clogged filters can restrict airflow, reducing the system's efficiency and potentially allowing contaminants to pass through.

Here's a guideline to keep your HVAC system running optimally:

- Schedule regular inspections of your HVAC system by a qualified technician.

- Change air filters every three to six months, or more often if the system is heavily used or if you have pets.

- Clean the coils and other components to prevent buildup that could hinder performance.

- Ensure proper sealing around filters to avoid air leakage.

- Monitor humidity levels as high humidity can promote mold growth, which standard filters might not handle effectively.

Following these maintenance tips ensures that your HVAC system continually operates at its best, providing cleaner air throughout your home (EPA et al., 2020).

In summary, selecting the right air filtration methods and maintaining them diligently is crucial for improving indoor air quality. HEPA filters, electrostatic air filters, and UV germicidal lamps each contribute uniquely to cleaner air, and when combined with regular maintenance, they form a comprehensive approach to enhancing the air we breathe.

Humidity Control Techniques

Humidity control in HVAC systems is a crucial aspect of maintaining optimal indoor air quality and preventing mold growth. Let's dive into some techniques and practices to help you keep your home's humidity at comfortable levels, and safeguard against those pesky molds.

High humidity levels can be particularly problematic. When the moisture in the air reaches high levels, it creates an ideal environment for mold to thrive.

Mold not only looks unsightly but can also exacerbate respiratory conditions such as asthma and allergies. According to the UGA Cooperative Extension, to prevent mold from becoming a permanent resident in your home, it's essential to manage moisture effectively (E. et al., n.d.).

One effective method to reduce moisture levels in your indoor environment is by using dehumidifiers. Dehumidifiers are appliances specifically designed to pull excess moisture from the air, creating a drier and more comfortable atmosphere. If you're considering using a dehumidifier, here are some tips to maximize its efficiency:

- Place the dehumidifier in areas with high humidity such as basements or bathrooms.
- Regularly clean and maintain the unit to ensure it operates efficiently.
- Ensure the dehumidifier has an auto-shutoff feature to avoid over-drying, which can also lead to discomfort.

Proper ventilation and air circulation are other critical factors in controlling indoor humidity levels and preventing condensation issues. Here's how you can enhance ventilation in your home:

- Use exhaust fans in kitchens and bathrooms to expel moist air generated during cooking and bathing.
- Install and regularly use ceiling fans to improve overall air circulation.
- Keep interior doors open to facilitate airflow between different rooms, especially if

you're experiencing dampness in certain areas.

- Use outdoor vented fans when possible to draw humid air outside rather than recirculating it indoors.

Monitoring and adjusting the humidity settings in your HVAC system is another layer of defense against excess moisture and related health issues. Modern HVAC systems often come equipped with built-in features to manage humidity. However, manual adjustments might be necessary to fine-tune the comfort level to your liking. Here's what you can do:

- Invest in a hygrometer, a device that measures humidity levels, to keep track of your indoor environment. Ideal levels should range between 30% and 50% relative humidity.

- Use programmable thermostats with humidity control capabilities. This allows you to set specific humidity levels for different times of the day, optimizing both comfort and energy usage.

- Regular maintenance of your HVAC system ensures filters are clean and free from blockages that can affect humidity control. Scheduling routine check-ups with a professional can help identify and rectify any inefficiencies.

By balancing these techniques—using dehumidifiers, improving ventilation, and meticulously managing your HVAC settings—you can achieve a comfortable indoor climate free from the risks associated with high humidity. It's about finding a harmony between human welfare and economic growth, where maintaining a healthy living space doesn't have to come at the expense of energy costs.

Let's consider a scenario: You've noticed that your basement consistently feels damp. The first step would be to measure the humidity. If you find the level exceeds 60%, it's time to take action.

Placing a dehumidifier in the basement will help mitigate this issue, alongside ensuring any potential leaks are fixed and adding insulation to cool surfaces. Proper ventilation systems would work to circulate air effectively, preventing stagnant air pockets where moisture likes to hide.

The key takeaway here is that maintaining appropriate humidity levels is vital not just for comfort, but for your health. Mold-related problems can escalate quickly if not addressed, leading to both structural damage and serious health concerns.

References

U.S. Environmental Protection Agency. (2014). Improving Indoor Air Quality. Learn the Issues, 2014(9), 27-26. Retrieved from https://www.epa.gov/indoor-air-quality-iaq/improving-indoor-air-quality.

U.S. Environmental Protection Agency. (2022). Guide to Air Cleaners in the Home. Other Policies and Guidance. https://www.epa.gov/indoor-air-quality-iaq/guide-air-cleaners-home

US EPA, OAR. (2014). Moisture Control, Part of Indoor Air Quality Design Tools for Schools. Overviews and Factsheets. Retrieved from https://www.epa.gov/iaq-schools/moisture-control-part-indoor-air-quality-design-tools-schools

Ogden, J. E., & Turner, P. R. (n.d.). Preventing Mold in Your Home. Retrieved from https://extension.uga.edu/publications/detail.html?number=C1047-1&title=preventing-mold-in-your-home

US EPA, OAR. (2014). The Inside Story: A Guide to Indoor Air Quality. Overviews and Factsheets. https://www.epa.gov/indoor-air-quality-iaq/inside-story-guide-indoor-air-quality

American Lung Association. (n.d.). Bringing fresh, outdoor air inside creates healthier indoor air.. Retrieved from https://www.lung.org/clean-air/indoor-air/protecting-from-air-pollution/ventilation

Farrell, M. H. J. (2020). 14 ways to reduce indoor air pollution. Consumer Reports. Retrieved from https://www.consumerreports.org/indoor-air-quality/ways-to-reduce-indoor-air-pollution/

US EPA, OAR. (2014). Introduction to Indoor Air Quality. Retrieved from https://www.epa.gov/indoor-air-quality-iaq/introduction-indoor-air-quality

Energy.gov. (n.d.). Moisture Control. Retrieved from https://www.energy.gov/energysaver/moisture-control

California Air Resources Board. (n.d.). Reducing your exposure to indoor air pollution. Retrieved from https://ww2.arb.ca.gov/resources/fact-sheets/reducing-your-exposure-indoor-air-pollution

ASHRAE. (n.d.). Learn more about Filtration / Disinfection. Retrieved from https://www.ashrae.org/technical-resources/filtration-disinfection

Environmental Protection Agency. (2020). Air Cleaners, HVAC Filters, and Coronavirus (COVID-19). Retrieved from https://www.epa.gov/indoor-air-quality-iaq/air-cleaners-hvac-filters-and-coronavirus-covid-19

9

HVAC Maintenance Practices

When you think about home comfort, it's easy to overlook the HVAC system silently humming in the background until something goes wrong. Yet, many homeowners don't realize that regular maintenance can significantly impact the longevity and efficiency of their HVAC equipment.

It's not just an afterthought; it's a vital practice that can prevent unexpected breakdowns and high repair costs.

The problem is clear: neglecting regular maintenance leads to inefficient performance and frequent, costly repairs. Over time, dust and debris accumulate in the system, causing everything from clogged filters and dirty coils to more severe mechanical issues.

Dirty air filters restrict airflow, forcing the system to work harder and consume more energy, which ultimately shortens its lifespan. Similarly, unchecked refrigerant levels can decrease cooling efficiency, making your air conditioner struggle to maintain comfortable indoor temperatures.

The consequences of such neglect are twofold—higher energy bills and an increased likelihood of sudden breakdowns during peak usage times.

Creating a Maintenance Schedule for HVAC Systems

Ensuring the longevity and efficient performance of your HVAC system isn't just a matter of convenience; it's an essential aspect of home maintenance that can prevent costly breakdowns, ensure comfort, and save on energy bills.

One of the most effective strategies to achieve this is by establishing a regular maintenance schedule.

Maintaining a regular schedule for HVAC system maintenance is essential for preventing breakdowns and extending its lifespan. By performing regular maintenance, you can detect and address small issues before they escalate into major problems. This will help keep your system running smoothly and efficiently for its entire lifespan.

According to ENERGY STAR®, annual pre-season check-ups are crucial—cooling systems should be checked in spring and heating systems in fall (ENERGY STAR®, n.d.).

Here's what you can do:

- Plan seasonal check-ups around time changes in spring and fall.
- Check thermostat settings to ensure the system keeps you comfortable when you're home and saves energy while you're away.
- Tighten all electrical connections and measure voltage and current on motors.
- Lubricate moving parts to reduce friction and electricity use.

Implementing these proactive measures not only enhances efficiency, but also prolongs the lifespan of critical components and enhances overall safety. Following these guidelines will help you prevent any surprise repairs and ensure a steady level of comfort indoors.

The maintenance schedule should include inspections, cleanings, and tune-ups to keep the system running efficiently and to identify any issues before they escalate.

For instance, cleaning the evaporator and condenser air conditioning coils can significantly improve the unit's cooling efficiency and prevent it from running longer than necessary, thereby saving energy and reducing wear and tear (ryno419, 2023).

It's important to regularly inspect the gas or oil connections, gas pressure, burner combustion, and heat exchanger functionality to make sure everything is running safely and efficiently. Following these steps is crucial for keeping your HVAC system in good condition.

One more important benefit of keeping a schedule is being able to keep track of past maintenance activities. This helps in planning for future HVAC service requirements more effectively. By maintaining thorough records, you can proactively address any potential issues and strategically prepare for required replacements or upgrades, avoiding unexpected expenses.

For example, knowing that your system's refrigerant level was low during the last check-up can prompt more frequent monitoring to avoid efficiency loss.

Using digital tools or reminder services can help automate maintenance schedule notifications for convenience. With the plethora of apps and digital calendars available today, it has never been easier to set and forget.

Set reminders for filter changes, scheduled professional inspections, and component cleanings. Many HVAC companies offer subscription-based services that include periodic check-ups and priority repair services, ensuring that you never miss critical maintenance.

Let's delve into how you can set up a practical and effective maintenance plan:

- Consult your HVAC system's manual or ask a professional to understand the recommended maintenance frequency.

- Break down tasks into manageable monthly, seasonal, and yearly activities. For example, inspect, clean, or change air filters every month.

- Schedule professional inspections twice a year—spring for cooling systems and fall for heating systems—to perform comprehensive checks and repairs (Saving$, 2023).

- Use digital reminders or subscription services offered by HVAC companies to streamline the process and ensure nothing is overlooked.

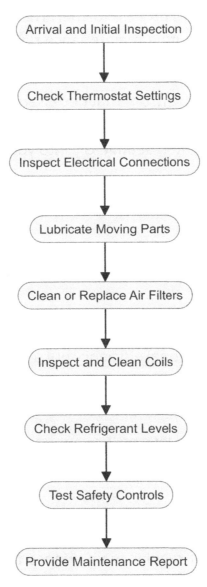

Typical HVAC Maintenance Checklist

By integrating these practices into your routine, maintaining HVAC systems becomes second nature, and the benefits will speak for themselves in terms of cost savings, improved efficiency, and extended equipment life.

You're not just safeguarding your investment, but also ensuring a comfortable living environment, reduced energy consumption, and a smaller carbon footprint. It's a win-win situation.

When setting up your maintenance plan, remember that attention to detail is vital. Small tasks like checking the condensate drain, tightening connections, and cleaning blower components can vastly improve system performance and delay the need for expensive replacements.

Think of it as investing in a long-term, comfortable, and efficient living experience.

Involving technology can simplify this process further. There are various digital tools and apps designed to send reminders for routine maintenance tasks, making it easier to adhere to a structured plan.

By automating these reminders, you ensure that necessary tasks aren't forgotten amidst daily life's hustle and bustle. This approach ensures that maintenance becomes less of a chore and more of a standard practice integrated seamlessly into your routine.

A practical tip for homeowners is to align HVAC maintenance schedules with other household tasks or seasonal changes.

For instance, daylight saving time adjustments can serve as a perfect reminder for bi-annual HVAC check-ups. This method leverages existing routines to establish new habits, making maintenance a no-brainer.

Ultimately, embedding an evidence-driven, systematic approach to HVAC maintenance into your daily life fosters not just operational efficiency but also peace of mind. Knowing that your home's HVAC system is in top-notch condition allows you to focus on other aspects of home improvement and personal well-being.

So, gear up, set those schedules, and let's ensure our homes remain the sanctuaries we need them to be.

Cleaning and Servicing HVAC Components

Let's delve into the significance of regular maintenance for your HVAC system and outline some actionable steps for cleaning and servicing its components.

First and foremost, keeping HVAC components like coils, filters, and vents clean is essential for improving system performance. Clean systems ensure that airflow remains unobstructed and reduce strain on the equipment, ultimately extending its lifespan (Advanced Energy, n.d.).

To achieve this, one should regularly check these parts to prevent dust build-up. Here are some steps you can take:

- Ensure you turn off the power before beginning any cleaning.

- Use a vacuum with a brush attachment to remove dust from the coils gently.

- Wash or replace filters as recommended by the manufacturer; typically, this needs to be done every one to three months.

- Make sure vents and ducts are free from obstructions and debris.

When cleaning these components, remember to be thorough yet gentle, avoiding any damage to critical parts. Consistent maintenance can substantially boost your system's efficiency, lowering energy use and minimizing the chances of unexpected breakdowns.

Next, it's important to address the broader benefits of cleaning and servicing your HVAC system, particularly when it comes to indoor air quality. Dust, debris, and microbial growth such as mold can accumulate within the system, significantly impacting the air you breathe. Clean components not only promote healthier air but also contribute to a more comfortable living environment (Kistler, 2022).

It's easy to overlook the direct connection between regular HVAC maintenance and health benefits, but it's an important aspect to consider. Just make sure to keep those filters and vents clean, and you'll be able to reduce the health risks that come with having poor indoor air quality.

Proper servicing goes beyond mere cleaning. It involves examining and greasing the moving components, inspecting electrical connections, and ensuring that everything is working properly. If you don't keep up with lubrication, it can cause more friction between the parts that move, and that can make them wear out faster. Electrical connections that haven't been checked can be dangerous and may lead to intermittent operation failures.

A comprehensive maintenance routine can prevent costly repairs down the road and help avoid sudden outages that leave you in discomfort during peak weather conditions.

If you're keen on tackling basic maintenance themselves, some tasks are pretty straightforward and quite manageable. Regularly replacing air filters, for example, is an easy task that doesn't require professional intervention.

Here's how you can do it effectively:

- Identify the type and size of the filter your HVAC system uses by consulting the manual or the current filter.

- Purchase the correct replacement filter from a home improvement store.

- Turn off the HVAC system.

- Open the filter compartment, remove the old filter, and insert the new one, ensuring it's correctly aligned according to the airflow direction indicated on the filter.

It's important to know when to reach out to professionals for more complex maintenance tasks, even though these DIY measures can be really helpful. When it comes to servicing components such as refrigerant levels, electrical wiring, and internal mechanisms, it's important to rely on certified technicians who have the specialized knowledge and tools to handle these tasks.

By adopting an evidence-driven approach, we realize that effective HVAC maintenance isn't just about prolonging the life of the equipment or reducing energy bills—though those are significant benefits. It's also about creating healthier living environments and preventing minor issues from escalating into major problems.

Checking and Replacing Air Filters

When it comes to ensuring the longevity of your HVAC system, paying attention to air filters is one of the most straightforward yet essential maintenance tasks. Air filters play a crucial role in maintaining indoor air quality and system efficiency by trapping dust, allergens, and pollutants from circulating in the air.

This simple mechanism helps not only in keeping your living space more comfortable but also in mitigating health issues related to poor indoor air quality, such as asthma or allergies.

Regular inspection and timely replacement of air filters are pivotal in preventing clogging, which can restrict airflow, strain the system, and lead to decreased efficiency. A clogged filter forces your HVAC system to work harder, consuming more energy and potentially decreasing its lifespan.

Here's what you can do to avoid these issues:

- Check the condition of your air filter every month. Remove the filter and hold it up to a light source to check for transparency. If light can't pass through easily, it's time for a change.

- Make it a habit to replace the filter at least once every three months. However, if your usage is high, you might need to perform this task more frequently.

Different types of air filters—fiberglass, pleated, HEPA—offer varying levels of filtration.

For instance, fiberglass filters are often the cheapest option but provide minimal filtration, mainly protecting the HVAC system itself rather than improving indoor air quality significantly. Pleated filters offer a higher level of filtration and can trap smaller particles like pollen and pet dander. HEPA filters, on the other hand, are top-of-the-line and can capture almost all airborne particles.

Selecting the right filter for your needs involves considering factors like the presence of pets, allergies within the household, and overall environmental conditions. The higher the MERV

(Minimum Efficiency Reporting Value) rating, the better the filter's ability to capture airborne particles, though it may also reduce airflow if not suited for your specific HVAC model.

Homeowners should be educated on the frequency of filter changes based on their usage patterns, pets, allergies, and environmental factors to ensure effective filtration.

- If you have pets, their dander can contribute to faster clogging of the filters. Thus, consider changing the filters every month.

- Families with multiple members or those who smoke indoors should also opt for monthly replacements to manage higher dust and pollutant levels.

- For homes located in windy areas or near construction sites, where dust and debris levels are higher, frequent checks and changes, possibly monthly, are advisable.

- If anyone in the house suffers from respiratory conditions, switching out filters often will help maintain a healthier living environment.

An often-overlooked aspect is the impact of a dirty filter on your HVAC system's operational costs. According to research, replacing a clogged filter with a clean one can lower your air conditioner's energy consumption by 5% to 15% (Homeowner Maintenance: Changing the HVAC Filter, n.d.).

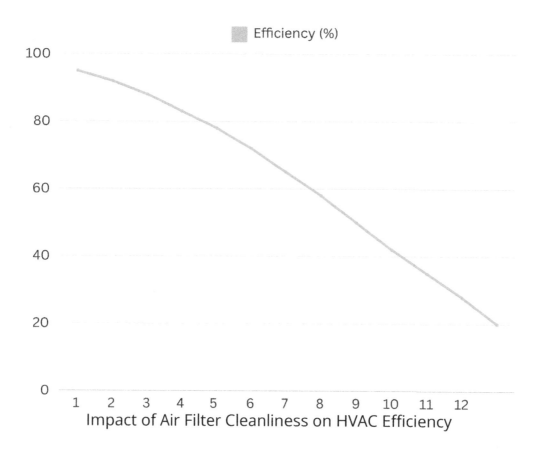

Impact of Air Filter Cleanliness on HVAC Efficiency

Not only does regular maintenance save you from costly repairs, but it also translates into noticeable savings on your utility bills.

Furthermore, by regularly changing your air filters, you can greatly enhance the air quality in your home. As time goes on, the air keeps flowing through those dirty filters, which means more and more contaminants end up in your living space. For people with chronic respiratory illnesses or allergies, this can make their symptoms worse.

By maintaining clean filters, you ensure that such particles are effectively trapped, creating a healthier indoor environment for everyone (Sweeney, 2024).

To seamlessly integrate this maintenance task into your routine:

- Set reminders on your phone or calendar to check the filters at regular intervals.

- Purchase filters in bulk so that you always have replacements on hand without having to make a last-minute run to the store.

Let's talk about replacing the filter itself, which is often simpler than many people think:

- First, turn off the HVAC unit to ensure safety.

- Locate the filter compartment, usually found near the return air duct or blower motor.
- Take out the old filter and carefully insert the new one, ensuring it's aligned properly.
- Turn the unit back on and observe its operation to verify correct installation.

Don't overlook the importance of cleaning up any residual dust around the filter housing before installing the new filter. This step ensures no loose particles get sucked into the system immediately after you've just changed the filter.

Additionally, make a note of the date you replaced the filter. Keeping a log can help identify trends or recurring issues, making troubleshooting easier if future problems arise.

Lastly, don't underestimate the benefits of consulting a professional for annual HVAC inspections. A trained technician can quickly identify underlying problems that might not be apparent during routine filter checks. They can also provide guidance on selecting the best filter type for your specific needs and ensure that the entire system operates harmoniously.

Seasonal Maintenance Tasks for Heating and Cooling Equipment

Seasonal maintenance of HVAC systems is not just a good practice; it's a necessity for anyone keen on ensuring reliable operation and comfort throughout the year.

Imagine braving a harsh winter evening only to find out that your heater has given up the ghost, or enduring a sweltering summer day with an air conditioner that simply refuses to cool. These scenarios are avoidable with a bit of routine attention and preparation.

By performing seasonal maintenance, you can make sure that your heating system is in top shape before the winter season hits. This will help prevent any unexpected heating failures and ensure that your system operates reliably. Here are the steps you can take to achieve this:

- Start by inspecting the heating elements for any apparent signs of wear or damage. Look for cracks, corrosion, or anything that seems out of place.
- Clean the elements thoroughly. Dust and debris can accumulate over time, impacting performance. Use a soft brush or vacuum to remove dust without damaging delicate components.
- Test the elements after cleaning to ensure they're working correctly. Turn on the system and observe if the heat distribution is uniform and effective.
- Check the surrounding area for any obstructions. Make sure there's ample space

around the heating unit to allow for proper airflow and functionality.

- Schedule professional servicing if needed. If any issues seem beyond your DIY capacity, reach out to a qualified technician to handle the intricate details (Energy.gov, n.d.).

Similarly, checking refrigerant levels, cleaning coils, and testing cooling functions ahead of the summer season help optimize cooling performance and energy efficiency. Here's how to get prepared:

- Begin by checking the refrigerant levels. Low refrigerant can severely affect the cooling capacity and efficiency of your unit. Make sure it's within the recommended levels and look for any potential leaks.

- Cleaning the coils is crucial. Dirty coils trap heat and make your system work harder, consuming more energy. Gently clean the evaporator and condenser coils using a coil cleaner or soapy water.

- Evaluate the cooling function by turning on the air conditioning and letting it run for a while. Listen for unusual noises and check if the cooling is consistent across all areas served by the system.

- Inspect the area around the outdoor unit. Remove any debris like leaves and twigs that might obstruct airflow. Trim back foliage at least two feet from the unit to keep it clear.

- Replace or clean the air filters regularly. Clogged filters impede airflow and reduce system efficiency. This simple step can lower energy consumption by 5% to 15% (Inflation today continues to keep prices of goods and services high, n.d.).

Seasonal maintenance tasks often involve evaluating thermostat settings, calibrating temperature controls, and testing safety features to ensure overall system functionality. Here's a set of steps to guide you through:

- Examine your thermostat settings. Ensure they align with your comfort needs. Programmable thermostats should be set to maintain comfortable temperatures when you're home and energy-saving ones when you're away.

- Calibrate the temperature controls. Use a thermometer to compare it with the readings on your thermostat. Adjust accordingly until they match.

- Test safety features like pressure relief valves and sensors that shut down the system in case of malfunctions. This ensures that your HVAC system operates safely and effectively.

- Review electrical connections and control sequences. Faulty wiring or misconfigured control sequences can pose hazards and reduce system lifespan. Tighten connections and measure voltage on motors to prevent unsafe operations.

- Check and clear condensate drains. Blocked drains can lead to water damage and affect indoor humidity levels. Pass a stiff wire through the drain channels to clear any buildup (ENERGY STAR®, n.d.).

One often overlooked yet critical aspect is filter maintenance. Filters trap dust, dirt, and other airborne particles, preventing them from circulating through your home. A dirty filter restricts airflow, making your system work harder, driving up energy use, and reducing the unit's lifespan.

Replace or clean filters every month or two during peak usage times. In dusty environments or homes with pets, it needs to happen more often (Inflation today continues to keep prices of goods and services high, n.d.).

Coils, another vital component, also require frequent attention. Over time, both evaporator and condenser coils accumulate dirt. A clean filter plays a significant role in slowing this process, but eventually, these coils will need cleaning too.

Yearly inspections and cleanings are usually enough, unless the outside surroundings are especially dusty or filled with leaves. Cleaning coils is essential for maximizing efficiency and prolonging their lifespan.

Cleaning Condenser Coils for Optimal Performance

Don't forget to take care of the exterior aspects of your HVAC system as well. Take the condenser unit outside, for example. It's subject to the whims of nature—dust, leaves, and other yard debris can quickly clog it up, obstructing airflow and causing it to overheat. Ensuring a clean, clear area around the unit is essential. Trim nearby plants and bushes and remove any debris promptly.

Additionally, don't overlook the fins on your air conditioner's coils. These aluminum fins can easily bend and block airflow. Investing in a fin comb—a specialized tool designed to straighten bent fins—can help maintain optimal airflow and efficiency (Energy.gov, n.d.).

Lastly, don't forget about the condensate drains. These drains play a key role in managing humidity levels within your home. Clogged drains can lead to increased moisture, fostering mold growth and causing water damage. Periodically passing a stiff wire through the drains helps maintain proper drainage.

Remember that having a professional inspection can provide reassurance that your system is up to any challenge. Your HVAC system, much like any other piece of machinery, thrives on attention and care. And when you take good care of it, it takes excellent care of you.

References

A Quality HVAC and Plumbing Services LLC. (2024). The role of filters in air conditioning systems. Journal of HVAC Systems. Retrieved from https://aqualityhvac.org/blog/the-role-of-filters-in-air-conditioning-systems/

Ryno419. (2023). Benefits of an HVAC Maintenance Membership. Hawk Heating. Retrieved from https://hawkair.org/discover-the-benefits-of-an-hvac-maintenance-membership/

Internachi. (n.d.). Homeowner maintenance: Changing the HVAC filter. Retrieved from https://www.nachi.org/change-hvac-filter.htm.

Bogleheads. (n.d.). Deducting real estate taxes on Sch F. Bogleheads. https://www.bogleheads.org/forum/viewtopic.php?t=406708

Kistler, S. (2022). Your respiratory health can rely on your HVAC system. Here's how to ensure higher-quality air in your home. Ohio State Health & Discovery. https://health.osu.edu/wellness/prevention/your-respiratory-health-can-rely-on-your-hvac-system

Travis Credit Union. (2023). Save money this summer by maintaining your HVAC. Travis Credit Union Blogs. Retrieved from https://www.traviscu.org/my-life/blogs/financial-wellness/march-2023/save-money-this-summer-by-maintaining-your-hvac/

Advanced Energy. (n.d.). The Importance of Properly Maintaining Your HVAC Systems. Retrieved from https://www.advancedenergy.org/news/the-importance-of-properly-maintaining-your-hvac-systems

Environmental Protection Agency. (n.d.). Maintenance checklist. Retrieved from https://www.energystar.gov/saveathome/heating-cooling/maintenance-checklist

Sweeney, A. (2024). The Importance of Regularly Changing Your Furnace Air Filter. A Message From The HVAC/R Department. Retrieved from https://ntinow.edu/the-importance-of-regularly-changing-your-furnace-air-filter/

U.S. Environmental Protection Agency. (2014). Should You Have the Air Ducts in Your Home Cleaned?. Indoor Air Quality (IAQ). Retrieved from https://www.epa.gov/indoor-air-quality-iaq/should-you-have-air-ducts-your-home-cleaned

ENERGY STAR®. (n.d.). Maintenance Checklist. Retrieved from https://www.energystar.gov/saveathome/heating-cooling/maintenance-checklist

U.S. Department of Energy. (n.d.). Maintaining Your Air Conditioner. Energy.gov. https://www.energy.gov/energysaver/maintaining-your-air-conditioner

10

Innovations in HVAC Technology

Picture walking into a home where the temperature is always perfect, no matter the season, and your energy bills are lower than you thought possible. This isn't science fiction—these are the benefits of modern HVAC systems powered by recent technological breakthroughs.

As homeowners and DIY enthusiasts just like you increasingly seek more control and efficiency in their living spaces, the latest innovations in HVAC technology offer smart solutions that promise to revolutionize how we manage indoor climates.

Traditional HVAC systems often operate at fixed settings, leading to unnecessary energy consumption and inconsistent comfort levels. These outdated methods are neither energy-efficient nor convenient, leaving homeowners with high utility bills and uneven temperature regulation.

The problem is further compounded by the lack of integration between various home systems, making it difficult to achieve a cohesive and intelligent home environment.

Smart HVAC Systems

Smart HVAC systems are revolutionizing the way we think about heating, ventilation, and air conditioning, offering a myriad of benefits that go beyond traditional setups. They use sensors and automation to seamlessly optimize energy usage and maintain indoor comfort levels—a significant leap forward for homeowners seeking efficiency and ease of use.

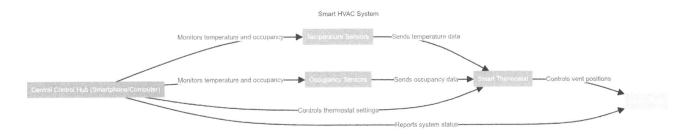

Smart HVAC System: A Network of Intelligent Devices

These systems use a variety of sensors—such as environmental sensors, occupancy sensors, and proximity sensors—to gather real-time data about your home's conditions and adjust settings accordingly (Vader, 2023).

Environmental sensors keep tabs on air quality and alert you if anything is awry, while occupancy and proximity sensors ensure that energy isn't wasted by adjusting temperatures based on whether someone is present in the room or how close they are to home.

Here's what you can do to achieve optimized energy usage and indoor comfort:

- Invest in a smart thermostat with built-in sensors.

- Ensure the environmental sensors are correctly placed around high-use areas like living rooms and kitchens.

- Configure occupancy and proximity sensors to accurately detect presence and movement within your home.

- Regularly check the system's app or interface to review energy consumption reports and make necessary tweaks.

The integration of smart technology also facilitates remote monitoring and control, making your life infinitely more convenient. Whether you're on a business trip or simply out for groceries, you can access your HVAC system through your smartphone or tablet. This allows you to monitor temperatures, adjust settings, and even receive alerts for any issues that may need your attention—all from the palm of your hand.

For instance, if an unexpected cold front moves in while you're away, you can turn up the heat without having to rush back home.

For those interested in maximizing the benefits of remote monitoring and control:

- Choose an HVAC system that supports mobile connectivity and has a user-friendly app.

- Set up your account and familiarize yourself with the app's features right from installation.

- Opt into notifications for maintenance reminders, temperature changes, and other important alerts.

- Experiment with automation rules, such as lowering the temperature during work hours and raising it before you return.

Another breakthrough feature of smart HVAC systems is their ability to provide predictive maintenance alerts. Traditional HVAC systems often leave homeowners scrambling when unexpected breakdowns occur, leading to costly repairs and downtime. Smart HVAC systems, however, analyze performance data and identify potential issues before they become significant problems.

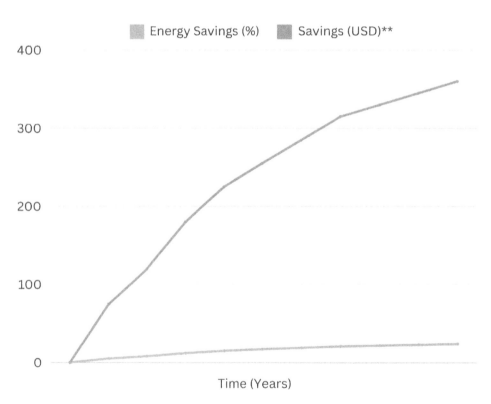

Predictive Maintenance: Maximizing Energy Savings Over Time

By sending alerts and maintenance reminders, these systems help you act proactively to prevent malfunctions and extend the life of your equipment (Murphy, 2023).

Here's how you can effectively use predictive maintenance:

- Activate maintenance alerts and set them to notify you well in advance of critical issues.

- Regularly review the system's diagnostic reports available through the app or interface.

- Schedule routine inspections and tune-ups based on the alerts you receive.

- Keep a log of maintenance activities and repairs for future reference and system optimization.

As we move towards an era of interconnected and intelligent buildings, adopting smart HVAC solutions aligns perfectly with this trend. These systems not only enhance individual comfort but also contribute to broader goals such as sustainability and energy conservation.

Intelligent HVAC systems integrate smoothly into smart homes, creating a cohesive environment where various technologies communicate and work together to maximize efficiency.

Think about a situation where your HVAC system collaborates with your smart lighting and security systems. When you leave your home, the lights dim, the doors lock, and the HVAC system adjusts to an energy-saving mode. Upon your return, everything reverts to your preferred settings, providing seamless comfort and security.

Implementing smart HVAC technologies can lead to substantial energy savings, enhanced comfort, and improved operational efficiency. It's not just a matter of convenience; it's a step towards smarter living and greater social responsibility.

So why wait? The future of HVAC is smart, and it's here now, ready to transform our lives one degree at a time.

Integration of IoT for Remote HVAC Control

IoT-enabled HVAC systems have revolutionized the way we manage and maintain indoor climates. By enabling real-time monitoring and control of temperature settings and air quality parameters, these smart systems ensure an optimal living environment with minimal manual intervention.

Imagine being able to check and adjust your home's temperature from your phone while you're out running errands or even while on vacation. This level of convenience can truly transform comfort in our everyday lives.

Real-time monitoring is particularly vital in maintaining a healthy indoor atmosphere. IoT devices can track air quality metrics such as humidity, particulate matter, and CO_2 levels, ensuring that the air you breathe remains clean and free from harmful pollutants.

This feature is particularly advantageous for families with kids, older individuals, or those who have respiratory problems. It effortlessly promotes a healthier environment in the living space, without any extra effort needed from the people living there.

Here's what you can do to leverage real-time monitoring and control effectively:

- Install smart thermostats and air quality sensors throughout your home.
- Connect these devices to a central system accessible through your smartphone or computer.
- Regularly check the metrics provided by these devices to stay informed about your home's air quality.
- Adjust settings as necessary to ensure optimal conditions year-round.

Remote access to HVAC controls brings another layer of convenience and efficiency. It allows homeowners and technicians to make adjustments and troubleshoot problems from virtually anywhere.

Consider a scenario where you're away from home during a sudden change in weather—remote access means you can preemptively adjust your heating or cooling settings to ensure comfort upon your return.

Moreover, this feature can be a lifesaver during vacations, allowing you to avoid energy waste by optimizing usage according to your absence.

For DIY enthusiasts or those looking to undertake simple maintenance tasks, remote access simplifies troubleshooting. If an issue arises, you or a technician can diagnose and potentially resolve the problem remotely, which can save both time and money.

Additionally, by connecting your HVAC system to an app or online platform, you can receive alerts about any irregularities, prompting quick action before minor issues escalate into significant repairs.

To make full use of remote access capabilities:

- Ensure that your smart HVAC devices are connected to a reliable network.
- Download relevant apps that allow you to control your HVAC system remotely.
- Familiarize yourself with the interface to quickly make adjustments when needed.
- Set up alerts for any anomalies to receive timely notifications.

IoT integration also significantly enhances data collection and analysis, leading to predictive maintenance and optimized energy usage. Sophisticated algorithms analyze data collected over time to identify patterns and predict potential failures before they occur.

By taking this anticipatory approach, you can proactively perform maintenance, which will help extend the lifespan of your equipment and prevent any unexpected breakdowns.

For instance, if your HVAC system's performance metrics indicate a gradual decline in efficiency, the system can alert you or your service provider to schedule a check-up or part replacement.

Predictive maintenance not only saves costs but also minimizes downtime, making it incredibly valuable for both residential and commercial applications.

You can know precisely when to replace an air filter or servicing a component before it deteriorates. This ensures that your HVAC system runs smoothly without interruptions and maintains high energy efficiency, which translates to lower utility bills.

To fully harness the power of predictive maintenance:

- Regularly review data insights from your HVAC system.
- Schedule routine check-ups based on predictive alerts rather than fixed intervals.
- Collaborate with your service provider to understand the patterns and metrics.
- Keep records of past maintenance activities and their outcomes for better future predictions.

The convergence of IoT with HVAC technology fosters a more responsive and adaptive approach to climate control. Traditional HVAC systems often adopt a one-size-fits-all approach, where manually adjusting vents and settings could be cumbersome and inefficient.

On the other hand, IoT-enabled systems can dynamically adjust based on current conditions and user preferences, leading to a comfortable and energy-efficient environment.

Smart vents equipped with temperature, motion, and proximity sensors can automatically open or close depending on the room's current state and occupancy. This ensures that energy is not wasted heating or cooling unoccupied spaces, and each room receives the optimal amount of air conditioning as needed.

Leveraging IoT capabilities in HVAC systems isn't just about immediate benefits like convenience and efficiency; it's about long-term gains through enhanced user experience, proactive maintenance, and data-driven decision-making.

As the demand for smarter, more responsive climate control solutions continues to grow, integrating IoT into your HVAC system is becoming less of a luxury and more of a necessity. Embracing IoT in HVAC isn't just about keeping up with trends—it's about paving the way for a more sustainable and comfortable future for all.

Recent Advances in Energy-Efficient HVAC Designs

Advancements in energy-efficient HVAC designs have shown impressive effects on sustainability. Let's explore the intricate details of these innovations and see how they not only achieve economic goals but also prioritize human well-being.

One of the key technological leaps in HVAC systems includes the integration of variable-speed compressors and advanced controls. These elements are revolutionary in that they allow HVAC units to operate at varying speeds depending on current needs, as opposed to traditional units that run at a constant speed regardless of demand.

This nuanced approach significantly optimizes performance, reducing unnecessary energy consumption. To incorporate these technologies effectively, consider:

- Upgrading to an HVAC system equipped with variable-speed compressors if your current unit runs continuously.

- Installing advanced control systems to monitor and adjust the operation of your HVAC system based on real-time data about temperature, humidity, and building occupancy.

- Regularly maintaining these modern components to ensure consistent performance and longevity.

Design features such as improved insulation, sealed ductwork, and zoned heating/cooling further enhance efficiency. Proper insulation maintains the desired temperature within a space, preventing unwanted heat transfer. Sealed ductwork ensures that air being heated or cooled by the HVAC system does not escape before reaching its intended destination, while zoning allows different areas of a building to be heated or cooled independently.

This is particularly beneficial for homes with multiple floors or large open spaces, where temperature preferences can vary greatly.

For homeowners looking to optimize their HVAC system, consider the following steps:

- Check existing insulation levels in your home and add more where necessary, focusing on attics, walls, and basements.

- Inspect ductwork for leaks using a smoke test or professional evaluation and seal any discovered gaps using mastic sealant or metal tape.

- Install dampers in your ductwork to create zones within your house, allowing you to control the temperature independently in different areas.

Advances in materials and construction techniques have also propelled the development of eco-friendly and sustainable HVAC solutions. Modern HVAC systems are increasingly crafted from recyclable materials, which can be disassembled and reused at the end of their life cycle, reducing overall environmental impact.

In addition, advancements in construction techniques have led to the creation of units that are more quiet and less obtrusive, which greatly enhances the overall indoor living experience. It's not necessary to follow strict guidelines when adopting these practices, but it's definitely a good idea to be aware of and prefer sustainable options when buying or upgrading HVAC systems.

Energy-efficient designs are fundamentally altering the landscape, providing benefits beyond mere cost savings. Lower operational costs, due to reduced energy consumption, translate directly into lower utility bills. More importantly, these designs contribute substantially to reducing carbon footprints, complementing broader environmental conservation efforts.

As homeowners, opting for energy-efficient systems is a small yet impactful step toward sustainability. Choosing ENERGY STAR®-rated appliances and regularly servicing your HVAC systems to maintain their efficiency are straightforward actions that support this goal.

The connection between economic growth and human welfare is evident when considering HVAC innovations. Modern systems not only save you money, but they also help create a healthier living environment by improving the quality of the air you breathe. Some of the latest HVAC units have these awesome HEPA filters that can actually trap all sorts of pollutants and allergens. This means that the air inside your home becomes a lot cleaner and safer, which is especially great news for people with respiratory problems.

Additionally, energy-efficient HVAC solutions often align with incentives and rebates offered by local, state, or national programs aimed at encouraging green initiatives. These financial incentives make upgrading to newer, more efficient systems an economically attractive option.

Emerging Trends in Sustainable HVAC Solutions

Exploring the latest advancements in HVAC technology involves taking a deep dive into sustainable solutions that aim to revolutionize the industry. Sustainable HVAC practices are gaining traction for their focus on energy conservation, renewable energy usation, and minimizing greenhouse gas emissions.

With climate change being a pressing concern, these advancements not only address environmental issues but also improve efficiency and reduce operating costs in the long run.

One primary objective of sustainable HVAC solutions is energy conservation. By implementing systems designed to use less power, we can significantly reduce our carbon footprint.

Take smart thermostats, for instance. They can adjust temperatures based on occupancy patterns and weather forecasting. This technology enables homeowners to optimize their energy usage without sacrificing comfort. The key here is automation that adapts to specific needs, cutting off unnecessary energy consumption seamlessly.

Another crucial aspect of sustainable HVAC practices is using renewable energy sources such as solar, geothermal, and wind power. Geothermal heat pumps, for example, make use of stable underground temperatures to heat and cool buildings efficiently.

This method drastically reduces reliance on fossil fuels and offers substantial energy savings over time. Plus, integrating these systems can be a game-changer for homes and commercial buildings looking to achieve net-zero energy status.

Here's what you can do to integrate green technologies like solar panels into your HVAC operations:

- Start with an energy audit to understand current consumption patterns.

- Consult professionals who specialize in eco-friendly HVAC systems to get tailored advice.

- Invest in high-efficiency solar panels and ensure they are strategically placed to maximize sunlight exposure.

- Pair solar panels with an electric heat pump to create a more robust and efficient system.

- Take advantage of government incentives and rebates available for renewable energy installations.

By following these steps, you can effectively enhance the sustainability and efficiency of your HVAC operations.

Sustainable HVAC solutions also prioritize indoor air quality, enhancing occupant comfort and holistic environmental performance within buildings. Innovations in air filtration systems, like

HEPA filters and UV germicidal lights, help to purify the air by removing contaminants and pathogens.

This not only contributes to a healthier living environment but also boosts productivity and overall well-being.

An often-overlooked factor in HVAC sustainability is the role of waste heat recovery systems. These systems capture excess heat generated in one area and redistribute it where needed.

For instance, waste heat from industrial processes or data centers can be repurposed to warm up office spaces or even provide hot water, thereby reducing the need for additional heating sources. It's a brilliant way to use what would otherwise be wasted energy.

The trends towards sustainability are driving innovation in HVAC design, installation, and maintenance practices. Manufacturers are continuously developing new, more efficient systems that require fewer resources.

One prime example is the rise of modular HVAC units, which can be customized and scaled according to specific building requirements. These units are easier to install and maintain, offering versatility and reducing the time and materials needed for traditional HVAC setups.

Moreover, the shift towards environmental-friendly refrigerants is another significant trend. Traditional refrigerants have been known to contribute to ozone depletion and global warming.

However, newer alternatives like hydrofluoroolefins (HFOs) have much lower global warming potential (GWP), making them a better choice for sustainable cooling solutions.

If you're considering retrofitting their existing HVAC systems, it's essential to evaluate the specific needs and constraints of their buildings. Retrofitting can range from simple upgrades, like installing programmable thermostats, to more complex modifications, such as integrating renewable energy sources or replacing old units with more energy-efficient models.

Regular maintenance and predictive diagnostics can also play a significant role in extending the lifespan and efficiency of your HVAC system.

Incorporating these sustainable practices requires a balanced approach that considers both economic growth and human welfare. It's about finding that sweet spot where technological advancement meets practical utility.

On a broader scale, the adoption of advanced HVAC technologies has significant consequences. Beyond individual homes, these systems can contribute to larger societal goals such as sustainability and energy conservation. As buildings become more interconnected and intelligent, the collective impact on reducing global energy demands and greenhouse gas emissions grows substantially.

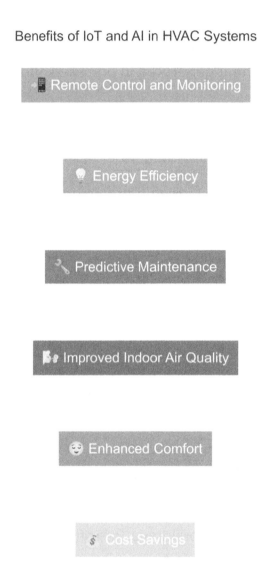

Benefits of IoT and AI in HVAC Systems

References

Asim, N., Badiei, M., Mohammad, M., Razali, H., Rajabi, A., Haw, L. C., & Ghazali, M. J. (2022). Sustainability of Heating, Ventilation and Air-Conditioning (HVAC) Systems in Buildings—An Overview. International Journal of Environmental Research and Public Health, 19(2), 1016. https://doi.org/10.3390/ijerph19021016

Vader, S. (2023). Are There Smart HVAC Systems?. Interactive College of Technology. Retrieved from https://www.ict.edu/news-events/are-there-smart-hvac-systems/

Asim, N., Badiei, M., Mohammad, M., Razali, H., Rajabi, A., & Haw, L. C. (2022). Sustainability of Heating, Ventilation and Air-Conditioning (HVAC) Systems in Buildings—An Overview. International Journal of Environmental Research and Public Health, 19(2), 1-16. https://doi.org/10.3390/ijerph19021016

ipl.org. (2020). Advantages and benefits of IoT in HVAC systems. Retrieved from https://www.ipl.org/essay/Advantages-And-Benefits-Of-IOT-In-HVAC-FCWKQY5Z26

Murphy, D. (2023). Advancing energy efficiency and sustainability in heating and cooling. ITI Technical College. Retrieved from https://iticollege.edu/blog/advancing-energy-efficiency-and-sustainability-in-heating-and-cooling/

Smith, J. (2024). Top trends in eco-friendly HVAC solutions. Sites At Penn State, 6(4). Retrieved from https://sites.psu.edu/socialtrends/2024/03/13/top-trends-in-eco-friendly-hvac-solutions/

DALL·E. (2024). Revolutionizing HVAC: The Impact of IoT on Air Duct Maintenance. DuctCleaning.org. https://www.ductcleaning.org/2024/02/09/revolutionizing-hvac-the-impact-of-iot-on-air-duct-maintenance/

ambitas.org. (2022). Use of IoT for Improved Indoor Air Quality in Facilities. Retrieved from https://www.ambitas.org/blog/use-of-iot-for-improved-indoor-air-quality-in-facilities

CMMigrants. (n.d.). The benefits of investing in a smart house HVAC system. Retrieved from https://cmmigrants.org/the-benefits-of-investing-in-a-smart-house-hvac-system/

Murphy, D. (2023). Smart Homes And HVAC Technicians Provide Ways To Remotely Control Thermostats And Other Devices. ITI Technical College. Retrieved from https://iticollege.edu/blog/how-hvac-technicians-are-adapting-to-changing-technologies/

Begley, R. (2024). What are the most eco-friendly HVAC systems?. The Beat by CBT Technology Institute. Retrieved from https://www.cbt.edu/blog/hvac/2024/what-are-the-most-eco-friendly-hvac-systems/

11
Heating Systems in HVAC

Heating systems are an essential part of any HVAC application, ensuring comfort and efficiency throughout the year. The core issue many homeowners face is selecting and maintaining an appropriate heating system that meets their needs without incurring exorbitant energy costs.

Forced air heating systems are quite prevalent, but they require meticulous upkeep. It's important to stay on top of filter changes and keep those ducts clean to avoid any build-up of dust. Another example is radiant heating, which offers silent operation and increased indoor air quality by not circulating dust or allergens.

But you really need to think about the type of flooring you choose so that heat can transfer efficiently. If homeowners don't have a good understanding of these principles, they may encounter issues such as inefficiencies and uneven temperatures in their living spaces.

Forced Air Heating Systems

Forced air heating systems have become incredibly popular in modern HVAC applications because of their high efficiency in both residential and commercial buildings. These systems are great at evenly distributing warm air, making every corner of your home or office feel cozy and inviting, especially during the winter months.

One of the main reasons for their popularity is their well-thought-out design, which guarantees consistent warmth in every room. This system operates by pulling air through return ducts into the furnace, where it gets heated before being distributed to different rooms through supply ducts and air registers. The continuous flow mechanism ensures that temperatures remain consistent and eliminates any concerns about hot or cold spots that may arise with alternative heating methods.

Moving deeper into the anatomy of forced air heating systems, it's essential to grasp the role played by components such as the furnace, ductwork, and air registers. The furnace serves as the heart of the system, generating the required heat.

Depending on the model, it could use electricity, natural gas, or propane as a fuel source. The ductwork, often sprawling across walls, ceilings, and floors, acts as the circulatory system, transporting the heated air to different parts of the building. Air registers then control the release of this warm air into individual rooms, making sure that each space achieves the desired temperature.

But it's not just about knowing their names and functions; it's about recognizing how they interact as a cohesive unit.

For instance, if the ductwork isn't properly sealed, even the most efficient furnace may struggle to heat your home effectively. Therefore, ensuring that each component is in optimal condition is vital to the overall performance of the system.

Regular maintenance of filters and ducts is crucial for keeping a forced air heating system in good condition. Over time, filters can get clogged with dust and debris, and ducts can accumulate buildup that hampers airflow and reduces efficiency. Here is what you can do to maintain the system:

- Regularly inspect and replace air filters. Dirty filters can restrict airflow, forcing your furnace to work harder and consume more energy.

- Schedule annual inspections to check the furnace's operation and ensure all safety measures are intact.

- Clean the supply and return air ducts annually to prevent dust and allergens from circulating within your home.

- Keep vents and registers clear of obstructions to facilitate unobstructed airflow.

In essence, diligent maintenance can significantly enhance the lifespan and efficiency of your heating system.

Upgrading to high-efficiency forced air heating systems can be a game-changer in terms of cost savings and energy efficiency. Modern high-efficiency furnaces can convert up to 98.5% of the fuel they consume into usable heat—an impressive leap from older models that operated at 56% to 70% efficiency.

Here is what you can do to achieve cost savings and improve energy efficiency:

- Consider upgrading to a high-efficiency furnace. These units typically come with advanced features like variable-speed blowers and electronic ignition that offer better performance and lower energy consumption.

- Evaluate the energy efficiency ratings (AFUE) of different furnaces before making a purchase. A higher AFUE rating indicates a more energy-efficient system.

- If your current system is old or frequently breaking down, compare the costs of retrofitting versus replacing it with a new high-efficiency model. Often, replacement is more economically viable in the long run.

- Incorporate programmable thermostats which optimize heating schedules based on occupancy patterns, further contributing to energy conservation and cost savings.

Regular maintenance and upgrading are not just about immediate benefits but also about long-term returns. Proper care and timely updates can extend the life of your HVAC system, making it a reliable component of your comfortable living or working environment for years to come.

Radiant Heating Technologies

Radiant heating systems use radiant heat transfer to warm objects and surfaces directly, creating a comfortable indoor environment.

In simpler terms, imagine feeling the warmth of the sun on your skin or standing close to a hot stove; that's how radiant heating works. Instead of warming the air around you, it warms the surfaces and objects in the room, which then radiate that warmth back to you. This method creates a cozy and evenly heated space, without the need for noisy fans or blowing dust around.

There are several types of radiant heating systems, each with distinct characteristics and benefits.

Types of Radiant Heating Systems

- **Hydronic (water-based) systems** circulate heated water through tubing laid under the floor, embedded in walls, or even placed in ceilings.

- **Electric systems** use heating cables or mats installed beneath the floor surface, which heats up when an electric current passes through them.

- **Infrared radiant panels**, often mounted on walls or ceilings, emit heat similar to the warmth you feel from the sun's rays.

These systems offer different installation options and energy efficiency levels, depending on what fits best for your home and lifestyle.

For anyone considering integrating radiant heating into their HVAC setup, understanding the design considerations is crucial. Here's what you need to think about:

- **Floor Construction:** The type of flooring can significantly impact the effectiveness and efficiency of radiant heating. Tile and concrete floors conduct heat well, distributing warmth efficiently across the room. On the other hand, carpeting and wood can insulate the heat, reducing system efficiency. If you must use insulating materials, keep them as thin and dense as possible to minimize heat loss.

- **Heat Output Requirements:** The size and insulation level of the room dictate the amount of heat required to maintain a comfortable temperature. Larger rooms or those with poor insulation will need more energy to heat effectively. To optimize performance, ensure your radiant heating system is properly sized to meet these requirements.

- **Zoning Controls:** Zoning allows for different parts of the house to be heated independently, enhancing both comfort and efficiency. With zoning controls, you can set varying temperatures in separate areas, ensuring you only heat the rooms you use most frequently. This can reduce energy consumption and lower utility bills.

To implement these design considerations effectively:

- **Consult a professional**: Whether you're installing hydronic, electric, or infrared systems, getting expert advice can prevent costly mistakes. Professionals can help determine the most suitable type of radiant heating for your specific needs and layout.

- **Evaluate your floor construction**: Choose flooring that optimizes heat transfer, like tile or concrete, especially if you're doing a wet installation where the heating elements are embedded in a slab.

- **Calculate heat output correctly**: Use tools or software to help calculate the heat loss in your space to ensure your system meets the necessary requirements.

- **Install zoning controls**: Work with a technician to install zoning valves or thermostats for customized control over different areas of your home. This way, you can manage your heating efficiently and comfortably.

Radiant heating systems are known for their quiet operation, consistent heat distribution, and potential energy savings over traditional forced-air systems. Unlike their noisy counterparts, radiant systems operate silently, providing a serene and comfortable living environment. The heat distribution is uniform, eliminating the common problem of cold spots found in many conventional heating methods.

One of the significant advantages of radiant heating is its energy-saving potential. Since radiant systems don't rely on air ducts, they avoid the typical energy losses associated with ductwork.

Additionally, radiant heating can be more efficient than baseboard heating and other forms of direct heat because it warms the room's surfaces rather than the air, which tends to escape quickly through doors and windows. This method not only conserves energy but also keeps your indoor environment healthier by not circulating allergens and dust particles.

If you have allergies or respiratory issues, consider this: Radiant heating doesn't blow air around your home, meaning fewer airborne particles, allergens, and pollutants get distributed. This can lead to a noticeable improvement in indoor air quality, making your living space more pleasant and healthier.

Integrating radiant heating with smart thermostats can further enhance efficiency. By programming your heating schedules, you can ensure the system operates only when needed, avoiding unnecessary energy consumption. Smart thermostats learn your preferences over time and adjust settings automatically, providing optimal comfort while saving money on your utility bills.

For homeowners looking to adopt eco-friendly heating solutions, radiant systems align well with sustainable practices. Hydronic systems, particularly, can be powered using renewable energy sources such as solar water heaters. By integrating these green technologies, you reduce your carbon footprint and contribute positively to the environment.

Geothermal Heating Options

When considering heating systems for your home or commercial space, geothermal heating options stand out not only for their efficiency but also for their minimal environmental impact. Let's delve deeper into how these systems work and the benefits they offer.

Geothermal heating systems use the stable temperature of the earth to provide heating and cooling through a series of underground loops that transfer heat to and from the ground. This process involves a network of pipes, often referred to as "ground loops," laid below the frost line where temperatures remain relatively constant throughout the year.

During winter, the system extracts heat from the ground and transfers it indoors. Conversely, in summer, the system reverses the process, pulling heat from the indoor air and discharging it into the ground. This method ensures consistent indoor temperatures year-round, moderated by nature's own thermostat.

Geothermal heating is incredibly energy efficient, which is a major advantage. According to the Environmental Protection Agency, geothermal ground source heat pump systems are among the most energy-efficient and environmentally clean space conditioning systems available. Approximately 70% of the energy used by these systems originates as renewable energy from the earth (Agency, n.d.).

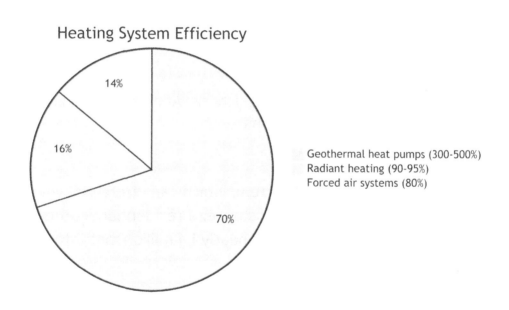

Efficiency Comparison of Heating Systems

High-efficiency geothermal systems can be up to 48% more efficient than traditional gas furnaces and even more so compared to oil furnaces (Agency, n.d.). But it's not just about efficiency; it's about sustainability and long-term savings.

While the initial installation costs of geothermal systems are higher than conventional HVAC systems, the investment pays off over time through reduced energy bills and lower maintenance costs. Geothermal heat pumps do not burn fossil fuels on-site, significantly reducing greenhouse gas emissions. They eliminate the risk of carbon monoxide within homes or buildings, contributing further to indoor air quality and safety.

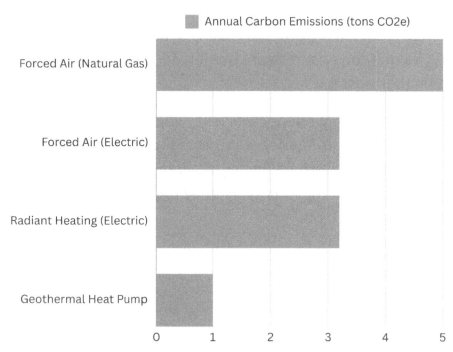

Environmental Impact of Heating Systems: Annual Carbon Emissions

Over the lifespan of a typical geothermal system, which can stretch 20 years or more, the environmental benefits are substantial. For instance, a residential geothermal heat pump system can reduce greenhouse gas emissions by nearly 1.1 million metric tons over its lifetime compared to conventional systems (Agency, n.d.).

Understanding the different types of geothermal systems is crucial in selecting the right one for your property. These systems generally fall into three categories: horizontal, vertical, and pond/lake loops.

- **Horizontal loops** are typically used for residences with ample land area, where pipes are laid out in shallow trenches.

- **Vertical loops** are ideal for buildings with limited land space, requiring deeper boreholes drilled vertically into the ground.

- **Pond or lake loops** use a body of water nearby. The coils are submerged, leveraging the water's thermal stability as a heat source or sink.

Each type has its unique advantages and limitations, depending on the geographical and structural characteristics of the site.

Here is what you can do to choose the best geothermal system for your needs:

- Assess the land area and availability of open spaces around your property.

- Consult with a geothermal HVAC professional to determine the suitability of your soil and rock formations for horizontal or vertical loop installations.

- If you have access to a nearby pond or lake, evaluate the feasibility of using a pond/lake loop system, ensuring that local regulations permit such installations.

- Consider the seasonal temperature variations and climate specifics of your region, which can affect the overall efficiency and effectiveness of the geothermal system.

Proper sizing, installation, and maintenance are essential to maximize the performance and longevity of geothermal heating systems.

To begin with, an accurate assessment of your heating and cooling requirements is necessary to size the heat pump appropriately. An undersized system might struggle to meet demands, while an oversized system could lead to inefficiencies and unnecessary costs.

Here is what you can do to ensure your geothermal system operates at its best:

- Have a qualified technician conduct a load calculation to determine the right size of the system required for your building.

- Ensure the installation is carried out by professionals familiar with geothermal technology to avoid common pitfalls like improper pipe layout or inadequate insulation.

- Regularly schedule maintenance checks to monitor the condition of ground loops, refrigerant levels, and the performance of the heat pump unit.

- Keep an eye on any unusual drops in efficiency and address them promptly by consulting with your service provider.

The environmental benefits of geothermal heating cannot be overstated. With the current focus on reducing carbon footprints and promoting sustainable practices, geothermal systems contribute significantly towards these goals.

Compared to traditional heating methods, geothermal systems produce fewer emissions throughout their operation. According to data from the Department of Energy, direct-use applications and geothermal heat pumps cause almost no negative effects on the environment;

instead, they help mitigate the reliance on energy sources that have adverse environmental impacts (EIA - Energy Information Administration, n.d.).

In fact, geothermal power plants emit 97% less sulfur compounds and about 99% less carbon dioxide than fossil fuel power plants of a similar size (EIA - Energy Information Administration, n.d.).

However, apart from just the numbers, incorporating geothermal technology into your HVAC project shows a dedication to safeguarding your property from unpredictable energy costs and promoting a healthier living environment. Although the initial cost may appear high, geothermal systems are a smart option for individuals who are thinking long-term due to their durability and dependability.

In addition to their environmental benefits, geothermal systems provide remarkable fiscal advantages. Lower energy consumption results in significant cost savings over time, easily outweighing the initial setup costs.

Furthermore, many regions offer incentives such as tax credits and rebates for installing renewable energy systems, further offsetting the expenses involved.

Ultimately, integrating geothermal heating into your HVAC strategy means harnessing a renewable resource that bridges the gap between personal responsibility and social duty, aligning individual choices with larger ecological benefits.

Heat Pump Systems

Heat pump systems are a marvel of modern HVAC technology, offering the unique ability to both heat and cool spaces by transferring heat between indoor and outdoor environments. This dual functionality ensures year-round comfort for homeowners, no matter the season. Imagine the practicality of a system that can seamlessly transition from warming your home during winter's bitter chill to cooling it during summer's sweltering heat.

The secret behind this versatility lies in their innovative design and efficient operation.

The types of heat pumps available today come in three main categories: air-source, ground-source (geothermal), and ductless mini split systems. Each type presents its own set of advantages tailored to various needs and preferences.

Air-source heat pumps are perhaps the most common and work by exchanging heat with the surrounding air. These systems have improved tremendously over the years, making them viable even in regions with harsh winters.

In contrast, geothermal heat pumps leverage the stable temperatures underground or within nearby water sources, achieving higher efficiencies, albeit with a steeper initial installation cost.

Meanwhile, ductless mini split systems offer tremendous flexibility, ideal for homes lacking existing ductwork or those wishing to avoid the complexity and expense of conventional duct installations. They don't just provide zoned heating and cooling; they also contribute to significant energy savings by transferring heat rather than generating it (Energy.gov, n.d.).

To truly appreciate the capability of heat pumps, it's crucial to delve into their inner workings. Heat pumps function through cycles of refrigerant that's compressed and expanded to draw heat from one location and release it at another. Understanding these mechanics is key for maintenance and troubleshooting.

Here is what you can do to achieve effective maintenance and troubleshooting:

- Regularly check the refrigerant levels to ensure they are adequate. Low refrigerant can significantly hamper the system's efficiency.

- Clean or replace air filters every few months to maintain optimal airflow and system performance.

- Inspect the outdoor unit for any obstructions, such as leaves or debris, that could block airflow and affect performance.

- Monitor the compressor's operation since it's the heart of the heat pump. Any unusual noises or vibrations could signal potential issues.

- For supplemental heating options, ensure backup heaters like electric resistance or gas furnaces are functioning correctly to maintain comfort on exceptionally cold days.

These steps not only help in maintaining the system's efficacy but also extend its lifespan, providing consistent and reliable comfort throughout the year.

One of the most compelling benefits of heat pump systems is their ability to reduce energy consumption significantly compared to traditional heating and cooling systems. Traditional systems, such as furnaces and boilers, generate heat through combustion or electric resistance, processes that are inherently energy intensive.

In contrast, heat pumps move heat rather than create it, resulting in substantial energy savings and lower utility bills for homeowners. According to ENERGY STAR®, certified mini split heat pumps can use up to 60% less energy than standard home electric radiators

(Ductless Heating & Cooling, n.d.). This impressive efficiency not only cuts costs but also reduces greenhouse gas emissions, contributing to environmental sustainability.

Let's talk about how heat pumps have improved over the years thanks to technological advancements. With the introduction of variable-speed motors and two-speed compressors, heat pumps have become much more adaptable and efficient. These innovations allow for airflow adjustments based on demand and optimize energy use, resulting in improved performance.

Features like desuperheaters, which recover waste heat for water heating, add another layer of efficiency, offering multi-functional benefits from a single system.

When considering a heat pump for your home, it's essential to evaluate your specific needs and circumstances to choose the right type. Whether you opt for an air-source, ground-source, or ductless mini split system, each has its niche where it excels.

If you're looking to upgrade an older system or install new equipment in a home without existing ducts, a ductless mini split might be the perfect solution. It avoids the need for extensive renovations and provides precise control over different zones in your home, enhancing comfort and efficiency.

In colder climates, opting for a cold climate ASHP (air-source heat pump) certified by programs like ENERGY STAR® ensures reliable performance even when temperatures plummet below freezing. These models use advanced compressors and refrigerants designed to maintain efficiency in severe conditions, ensuring that your home remains warm and cozy without a significant spike in energy consumption.

Additionally, many utilities and federal programs offer incentives and rebates for installing energy-efficient heat pumps, making them a financially savvy choice for long-term savings.

For instance, you can take advantage of federal tax credits of up to $2,000 for purchasing and installing qualified systems (Air-Source Heat Pumps, n.d.), further offsetting the initial investment.

Investing in a heat pump system is not just about enjoying immediate comfort. It's a forward-thinking decision that prioritizes long-term sustainability and cost-effectiveness. By choosing a system that aligns with your home's specific requirements, you can maximize both efficiency and comfort, transforming how you experience heating and cooling.

And as technology evolves, staying updated on best practices will enable you to keep your home warm and inviting, reflecting a commitment to both personal well-being and broader sustainability.

References

Energy, U. (n.d.). Heat Pump Systems. Energy.gov. Retrieved from https://www.energy.gov/energysaver/heat-pump-systems

American Society of Home Inspectors. (2023). What Is Forced-Air Heating? (2023 Guide). ASHI. Retrieved from https://www.homeinspector.org/consumers/hvac/what-is-forced-air-heating

Energy.gov. (n.d.). Radiant Heating. Retrieved from https://www.energy.gov/energysaver/radiant-heating

ENERGY STAR®. (n.d.). Air-Source Heat Pumps. Retrieved from https://www.energystar.gov/products/air_source_heat_pumps

Environmental Protection Agency. (n.d.). Efficiency and Environmental Benefits. Retrieved from https://www.energyhomes.org/renewable-technology/geoefficiency.html

Tulsa Welding School. (2023). What is Radiant Heating?. Tulsa Welding School. https://www.tws.edu/blog/skilled-trades/what-is-radiant-heating/

Environmental Protection Agency. (n.d.). Ductless Heating & Cooling. Retrieved from https://www.energystar.gov/products/ductless_heating_cooling

Energy Department. (n.d.). Furnaces and Boilers. Energy.gov. https://www.energy.gov/energysaver/furnaces-and-boilers

Department of Energy. (n.d.). Geothermal FAQs. https://www.energy.gov/eere/geothermal/geothermal-faqs

Plastic Pipe and Fittings Association. (-). Radiant Heating & Cooling Systems. Retrieved from https://www.plasticpipe.org/BuildingConstruction/BuildingConstruction/-Applications-/Radiant-HeatingCooling-Systems.aspx.

Energy Information Administration. (n.d.). Environmental issues and benefits related to use of geothermal energy. Energy Information Administration - EIA. Retrieved from https://www.eia.gov/energyexplained/geothermal/geothermal-energy-and-the-environment.php

12

Cooling Systems in HVAC

Many homeowners find themselves puzzled by the complexities of HVAC cooling systems. With options ranging from central air conditioning units to ductless mini splits and even evaporative coolers, understanding the benefits and limitations of each can seem daunting.

Take central air systems, for example. They're great for cooling large areas, but they do come with some downsides. You'll need to have extensive ductwork installed and make sure to keep up with regular maintenance. On the other hand, ductless mini split systems provide zoned cooling without the need for ducts but come with higher upfront costs.

Each system has its unique set of challenges and advantages, making it crucial to choose wisely based on your specific needs and conditions.By understanding the nuances of each system, you'll be better equipped to make informed decisions about maintaining or upgrading your cooling solution at home or in your business, ensuring year-round comfort and energy efficiency.

Understanding the Fundamentals of Central Air Conditioning Systems

As we dive into the fundamentals of central air conditioning systems, it's clear that these are the go-to choice for keeping many modern homes cool and comfortable. Central air conditioning systems provide consistent temperature control throughout an entire property, ensuring a uniform climate in every room.

This consistency forms one of their biggest advantages over other cooling methods that may only target specific areas.

Wouldn't it be a dream to walk into your home on a scorching summer day to feel a refreshing blast of cool air uniformly spread across every corner? That's the magic central air conditioning systems bring. These systems work by circulating cool air through a series of supply and return ducts.

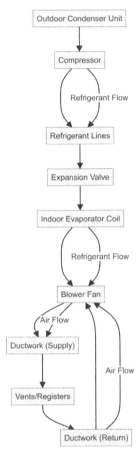

Central Air Conditioning System: Components and Airflow

The conditioned air moves through the supply ducts and registers—openings in floors, walls, or ceilings—where it silently and efficiently cools each room. Once the air warms up, it flows back through return ducts, gets reconditioned, and the cycle starts anew.

Beyond simply providing comfort, these systems contribute to a home's overall livability by reducing humidity levels and enhancing indoor air quality.

However, achieving this efficiency isn't automatic—it requires diligent maintenance.

Proper upkeep of central air conditioning units significantly impacts their performance and longevity. Regular filter changes and professional tune-ups are essential tasks that should not be overlooked. When filters get dirty, they can block the airflow, which makes your system work harder and use up more energy. This can result in less cooling being delivered.

To keep your central air conditioning unit performing optimally, here are some guidelines:

- Replace or clean your air conditioner's filters every month or two during the cooling season. If the unit is constantly running or exposed to dusty conditions or if you have pets, more frequent attention might be necessary.

- Check the evaporator coil annually and clean it as required to maintain its heat-absorbing capacity.

- Ensure that dirt and debris around the outdoor condenser unit are minimized, removing any foliage or debris that could block airflow (Energy.gov, n.d.).

These steps help avert common issues like clogged filters, dirty coils, and blocked airflow—all of which substantially degrade the system's effectiveness.

Additionally, using programmable thermostats in central air systems can help maximize both energy efficiency and comfort levels. By automatically adjusting the temperature based on your schedule, these devices ensure the system isn't cooling the house unnecessarily when no one is home, thereby reducing strain on the unit and saving energy.

Let's explore how to make the most out of programmable thermostats:

- Set the thermostat to higher temperatures when you're away from home. This small change can save between 5% and 15% on cooling energy.

- Use the "hold" or "vacation" setting when you'll be away for several days to prevent the system from cooling an empty home.

- During the summer, aim to set your thermostat to 78 degrees Fahrenheit when you're home and need cooling, but higher when you're away.

Programmable thermostats are a simple, yet impactful addition to your HVAC setup, offering both convenience and cost savings. But there's another critical area that demands our attention—ductwork.

Regular inspection of ductwork and seals is essential for maintaining efficient airflow and preventing energy waste. Leaky ducts can account for significant energy losses, causing your system to work harder than necessary to achieve desired cooling levels.

Over time, this not only raises your utility bills but also shortens the life span of your HVAC equipment.

To ensure your ductwork remains in top shape, consider the following practices:

- Inspect ducts for leaks and seal them with mastic sealant or metal tape.

- Insulate ducts that run through unconditioned spaces, such as attics or garages, to prevent energy loss.

- Schedule a professional inspection and cleaning of your duct system regularly, particularly if you notice dust accumulating in your home or if someone in the household has allergies.

Making sure that ducts are properly sealed and insulated is important not only for immediate comfort, but also for the long-term efficiency of your system and the overall energy health of your home.

By following these practices, you can fully enjoy the benefits of their central air systems without wasting energy or experiencing unexpected breakdowns.

Here's to cooler summers and smarter energy use!

Exploring the Benefits and Applications of Ductless Mini Split Systems

Ductless mini split systems are a fantastic option for zoned cooling solutions, allowing homeowners to control the temperature in specific areas of a home or building. This is particularly beneficial for those who have different cooling needs in various rooms.

For instance, perhaps you like your bedroom cooler at night while preferring a milder temperature in the living room during the day. With ductless mini splits, you can achieve this kind of personalized comfort effortlessly.

Another advantage of ductless minisplit systems is their installation process, which is less invasive compared to traditional ducted systems. If you're considering a retrofit project or adding an extension to your home, ductless mini splits could be the perfect fit. Installing these systems involves minimal structural modifications and typically requires just a small hole for connecting the indoor and outdoor units.

This feature not only helps cut down on labor expenses, but also simplifies the process of incorporating modern HVAC solutions into older structures without causing any disturbance to the existing architecture.

For installing ductless mini splits in these situations:

- Identify the best locations for the indoor air-handling units, keeping accessibility and aesthetic considerations in mind.

- Ensure that the outdoor unit is placed in a location with unobstructed airflow.

- Simplify the installation by running refrigerant lines through less visible spaces like closets or basements.

- Choose a professional installer experienced in handling ductless systems to avoid any potential issues.

Energy efficiency is another compelling reason to consider ductless minisplits. Unlike traditional systems that rely on extensive ductwork, these systems deliver cooled air directly to the zones where it's needed, significantly reducing energy loss. Traditional HVAC systems can lose up to 30% of energy through duct leaks, especially if the ducts run through unconditioned spaces like attics.

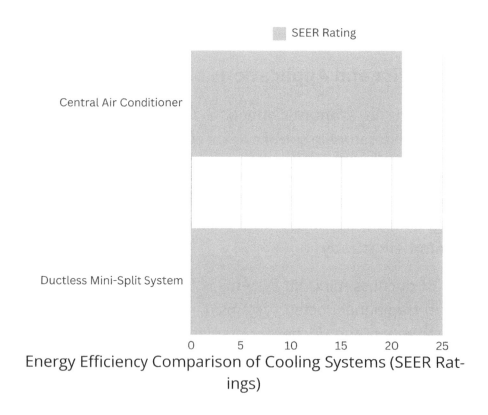

Energy Efficiency Comparison of Cooling Systems (SEER Ratings)

By contrast, ductless systems avoid this inefficiency altogether, making them a more eco-friendly and cost-effective option over time (Ductless Heating & Cooling, n.d.).

Regular maintenance is key to ensuring optimal performance and longevity of your ductless minisplit system.

One critical task is the regular cleaning of filters. Over time, dust and debris can accumulate on the filters, impeding airflow and causing the system to work harder, which can lead to higher energy bills and reduced indoor air quality.

To maintain the effectiveness and efficiency of your ductless minisplit:

- Check your filters monthly and clean them when they appear dirty. In dusty environments or homes with pets, more frequent cleaning might be necessary.

- Follow the manufacturer's instructions for removing and cleaning the filters.

- Consider setting reminders on your calendar or smartphone so you don't forget this crucial step.

Providing flexibility and efficient cooling options, ductless mini split systems shine in situations where traditional ductwork isn't feasible. They're perfect for homes with non-ducted heating systems like hydronic panels or space heaters and can seamlessly blend into contemporary interior designs, thanks to their sleek, high-tech look (Energy.gov, n.d.).

The versatility extends to commercial spaces too, where different rooms may have varying cooling requirements depending on occupancy levels and usage patterns.

One potential downside of ductless minisplits is the upfront cost, which can be higher compared to traditional HVAC systems or window units.

However, this initial expense is often offset by lower operating costs due to improved energy efficiency. Moreover, many regions offer rebates and incentives for installing ENERGY STAR® certified mini split systems, which can further reduce the financial burden.

It's worth consulting local utility companies or checking government websites to explore available options.

You should ensure that each indoor unit is properly sized and placed in order to fully maximize the advantages of ductless mini split systems. A larger unit can cause short cycling, where the system turns on and off frequently, wasting energy and failing to maintain a consistent temperature.

Conversely, an undersized unit will struggle to meet cooling demands, leading to overwork and potential breakdowns. Here's how to ensure proper sizing:

- Work with a certified HVAC contractor who uses tools like Manual J calculations, designed to determine the precise heating and cooling load requirements for your space.

- Consider factors such as the size of the area, insulation quality, number of occupants, and typical usage patterns during the assessment.

- Discuss the suitability of your selected unit for both summer cooling and winter heating if it offers both functions.

Although the indoor units may not be as visually appealing as a centralized system, the advantages usually outweigh this minor issue. The various mounting options—wall-mounted, floor-standing, or ceiling-recessed—provide flexibility to blend the units into your decor seamlessly.

Additionally, many models come with remote controls, adding convenience to functionality.

Another consideration is the placement of condensate drainage near the outdoor unit. Since mini splits extract moisture from the air, they need an effective way to expel the collected water. Ensuring there's a suitable drainage solution nearby can prevent water-related issues and keep the system running smoothly.

Although the initial investment may be higher, long-term savings, coupled with potential rebates, make them a financially sound choice. By working with a knowledgeable contractor and carefully considering unit placement and sizing, homeowners and businesses alike can enjoy the myriad benefits these systems offer.

Understanding the Functionality and Benefits of Evaporative Coolers for Efficient Cooling

Evaporative coolers, often colloquially referred to as swamp coolers, leverage the natural process of evaporation to provide a cost-effective and energy-efficient cooling solution for indoor spaces.

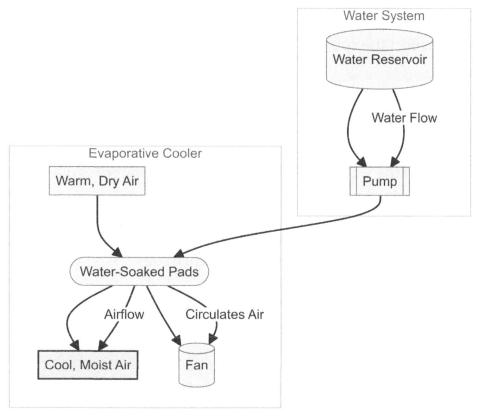

Evaporative Cooling: A Natural and Energy-Efficient Process

Unlike traditional air conditioning systems that rely on refrigerants, these coolers simply draw warm, dry air through water-soaked pads, facilitating the evaporation of water, which in turn cools the surrounding air before it is circulated indoors. This method not only cools the air but also adds humidity, making the environment feel more comfortable (AHRI, 2024).

One of the significant advantages of evaporative coolers over conventional air conditioners is their lower energy consumption. Traditional air conditioning units can be heavy power guzzlers due to the mechanical operations involved in refrigerant cycles.

In contrast, evaporative coolers require a fraction of that energy, thus reducing operational costs and the environmental footprint. Studies indicate that two-stage evaporative coolers can slash energy consumption by as much as 60% to 75% compared to standard air conditioning systems (Energy.gov, n.d.).

This makes them an appealing option for eco-conscious homeowners and businesses looking to cut down on their energy bills.

However, like all HVAC systems, evaporative coolers necessitate regular upkeep to ensure they perform optimally and last longer. A critical aspect of this maintenance involves keeping the water pads and reservoirs clean to prevent the growth of mold and bacteria.

Here's what you can do to achieve effective maintenance:

- Frequently inspect the cooler's water pads and filters for any signs of wear or mineral build-up.

- Regularly clean or replace the pads as per the manufacturer's instructions—typically, this should be done at least twice during the cooling season.

- Empty and clean the water reservoir with soap and water regularly to avoid sediment accumulation and bacterial growth.

- Ensure the water pump and any moving parts are functioning correctly and haven't succumbed to mineral deposits or mechanical issues.

- During periods of inactivity, especially when switching off the unit for the season, make sure to drain the system completely and allow it to dry out to inhibit mold development.

It's worth noting that evaporative coolers work best in regions with low humidity. In dry and desert-like areas, like the southwestern United States, the air can easily soak up moisture from the pads, which makes the cooling process very efficient.

For instance, in cities like Phoenix, Arizona, where the climate conditions are ideal, evaporative coolers can lower air temperature significantly, rendering a comfortable living environment even during peak summer months (Building America Solution Center, 2022).

Conversely, in regions with higher humidity levels, the effectiveness of these coolers diminishes since the air is already near saturation and can't support further evaporation efficiently.

The sustainability credentials of evaporative coolers extend beyond just reduced energy use. Given their reliance on water rather than refrigerants, they pose minimal risks concerning ozone depletion and greenhouse gas emissions.

In addition, the initial installation expenses for evaporative coolers are usually only half of what it would cost to install a central air conditioning system. This makes them a financially appealing choice. This affordability opens up air cooling solutions to a broader audience who might otherwise find conventional air conditioning prohibitively expensive (Energy Saver, n.d.).

However, there are certain limitations and considerations homeowners need to account for. While the addition of humidity may be beneficial in dry climates, it can be a drawback where high humidity is already an issue.

In addition, it is important to have open windows or vents to allow evaporative coolers to continuously exchange indoor air with fresh outdoor air. It's important to find the right balance of open windows or vents. If there's not enough opening, the indoor humidity will increase. On the other hand, if there's too much opening, the cooling might not be sufficient.

Based on the points mentioned, it's clear that evaporative coolers provide a practical and sometimes better option compared to traditional air conditioning systems in certain environments. These devices are great for dry climates because they can both cool and humidify the air. Not only do they offer the potential for significant energy savings, but their environmentally friendly nature also makes them even more appealing.

Nevertheless, prospective users must commit to regular maintenance and be mindful of climatic suitability to harness the full benefits of these systems.By emphasizing diligent care and strategic usage tailored to local climate conditions, these systems can deliver consistent, eco-friendly comfort throughout the warmer months.

Exploring the Use of Chilled Water Systems in Commercial HVAC Settings

When discussing the diverse range of cooling technologies in HVAC systems, chilled water systems stand out as a remarkable solution. These systems use cold water circulated through coils to absorb heat from indoor spaces, making them highly efficient for large commercial buildings.

Imagine water at around 40° F moving through a network of pipes and heat exchangers, working tirelessly to maintain comfortable temperatures within expansive office complexes or shopping malls.

Scalability is a major benefit of chilled water systems. Whether it's a small section of a building or multiple zones across various floors, these systems can handle it all. They are designed to meet the cooling needs of distinct areas with different temperature requirements simultaneously.

The cool water absorbs heat in one area, moves through a return loop, and is re-cooled at a central plant before being recirculated. This flexibility makes chilled water systems ideal for settings where precise climate control is necessary.

Now, while these systems are sophisticated and effective, they do require a certain level of diligence to keep them operating at peak efficiency. Regular monitoring of water temperatures and pressure levels is crucial.

If you're a homeowner or a DIY enthusiast wanting to understand more about maintaining an HVAC system, here's something straightforward: consistently check your system's data logs or install sensors that provide real-time updates. Ensuring the system components like sensors and gauges work correctly will prevent unexpected downtimes and optimize performance.

Insulation and sealing of the chilled water pipes are extremely important. If the insulation is not up to par, the system can experience major energy losses. This not only impacts its efficiency but also leads to higher operational expenses. Things like condensation and thermal loss can degrade system components over time, leading to costly repairs or replacements.

If you're looking to implement this at home, consider these guidelines to achieve optimal insulation:

- **Select the correct insulation material:** Foam glass, elastomeric foam, and fiberglass wrapped in vapor barriers are excellent choices.

- **Ensure a continuous vapor barrier:** A break in the barrier allows moisture to seep in, creating potential for mold growth and decreased insulation effectiveness.

- **Seal joints and seams effectively:** Use high-quality adhesives and tape to ensure no gaps remain where conditioned air could escape.

- **Hire skilled professionals:** When in doubt, seek out experienced contractors who specialize in mechanical insulation.

By effectively managing insulation, not only do you maintain the integrity of your chilled-water system, but you also preserve its efficiency, translating into substantial energy savings over time.

Chilled water district cooling is a more advanced version of chilled water systems, specifically designed for campus environments. Centralized chilling plants generate and distribute cold water through underground insulated pipes, which helps improve energy efficiency and reduce costs (Second Nature, n.d.).

Retrofitting existing systems with Variable Speed Drive (VSD) compressors can further improve efficiency by adjusting compressor speeds based on the cooling demand. Lowering the temperature of the chilled water and optimizing flow rates are other methods to enhance system efficiency (Second Nature, n.d.).

When it comes to maintaining a chilled-water system, it's important to remember that it's not just about the central plant. Filters, pumps, and other ancillary equipment are also crucial for its operation. Efficient filtration systems are key to keeping particulate matter away, which in turn prevents component failure and guarantees long-term reliability(Insulation Outlook Magazine, 2017).

Furthermore, it is crucial to choose suitable materials and ensure impeccable installation practices when dealing with construction projects and upgrades that involve chilled-water systems. Customized insulation solutions can greatly benefit various components like ductwork, control valves, and heat exchangers.

Elastomeric insulation, known for its flexibility and permeability characteristics, stands out as a material of choice, especially for equipment requiring maintenance access and durability against tough environments (Insulation Outlook Magazine, 2017).

Moreover, integrating cooling towers into the chilled water system can offer dual benefits. Not only do they expel waste heat efficiently, but they also serve as reservoirs, storing cold water when electricity is cheaper or cleaner.

Operators have the option to turn off chillers and use stored cold water during peak demand times. This helps to decrease dependence on potentially less clean grid energy (Second Nature, n.d.). Cooling towers themselves benefit from modern design improvements like multiple tower operations, which allow for better heat transfer and reduced energy usage.

It's clear that while chilled water systems bring immense benefits in terms of energy efficiency and scalability, they are not without their challenges. Each system requires a specific setup, diligent monitoring, and periodic maintenance.

Choosing the right HVAC system involves considering various factors, including the specific needs of your home or building, local climate conditions, and long-term maintenance requirements. To achieve lasting comfort, it's important to understand and effectively maintain the chosen solution, as each system has its own unique benefits.

Embracing new innovations and honing maintenance skills will not only improve system efficiency but also contribute to more sustainable and cost-effective cooling solutions. As we continue exploring HVAC systems, let's remain curious and proactive, ensuring our living spaces are as comfortable and energy efficient as possible.

References

Johnson, L. (2022). Basic HVAC Troubleshooting Tips Every Homeowner Should Know. My Green Montgomery. https://mygreenmontgomery.org/2022/basic-hvac-troubleshooting-tips-every-homeowner-should-know/

Energy.gov. (n.d.). Evaporative Coolers. https://www.energy.gov/energysaver/evaporative-coolers

Pacific Northwest National Laboratory. (n.d.). Evaporative cooling systems. Retrieved from https://basc.pnnl.gov/resource-guides/evaporative-cooling-systems

AHRInet. (n.d.). How swamp coolers work. https://www.ahrinet.org/scholarships-education/education/homeowners/how-things-work/swamp-coolers

Energy.gov. (n.d.). Maintaining Your Air Conditioner. Retrieved from https://www.energy.gov/energysaver/maintaining-your-air-conditioner

Second Nature. (n.d.). Chilled Water District Cooling - Second Nature. Retrieved from https://secondnature.org/solutions-center/chilled-water-district-cooling/

ENERGY STAR®. (n.d.). Ductless Heating & Cooling. https://www.energystar.gov/products/ductless_heating_cooling

Energy Department. (n.d.). Ductless Mini split Air Conditioners. Energy.gov. Retrieved from https://www.energy.gov/energysaver/ductless-mini split-air-conditioners

Insulation Outlook Magazine. (2017). Chill Out: Maintaining Integrity of Chilled-Water Systems. Insulation Outlook Magazine. Retrieved from https://insulation.org/io/articles/chill-out-maintaining-integrity-of-chilled-water-systems/

Certified Commercial Property Inspectors Association. (2019). Central Plant Systems and Chilled Water Systems. Retrieved from https://ccpia.org/central-plant-systems-and-chilled-water-systems/

Energy.gov. (n.d.). Ductless Mini split Heat Pumps. Retrieved from https://www.energy.gov/energysaver/ductless-mini split-heat-pumps

Consumer Reports. (2017). How to Maintain Your Central Air Conditioning Units. Consumer Reports. Retrieved from https://www.consumerreports.org/central-air-conditioning/how-to-maintain-central-air-conditioning-units/

13
HVAC System Design Principles

Understanding how to design an HVAC system is akin to mastering a complex yet rewarding puzzle, where each piece must fit perfectly to create a harmonious whole.

At its core, designing an efficient HVAC system transcends the mere alignment of ducts and vents; it's about orchestrating a symphony of comfort, energy efficiency, and reliability.

One of the key challenges in HVAC system design is accurately calculating the load required for sizing equipment.

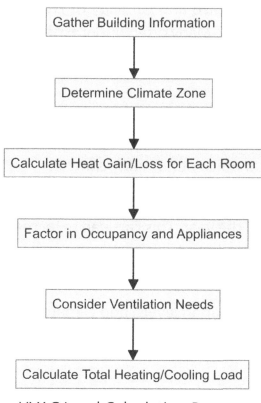

HVAC Load Calculation Process

Incorrect load calculations often lead to inefficiencies such as over-sizing or under-sizing units. An oversized unit tends to cycle on and off frequently, which can lead to a shorter lifespan and less effective humidity control. This not only increases energy consumption but also causes discomfort.

On the other hand, an undersized system struggles to maintain desired temperatures during peak seasons, leading to increased operational costs and potentially inadequate climate control.

A home in a hot, humid region like Phoenix will require different load specifications compared to one in a cooler, wetter climate like Seattle. Factors such as building size, insulation levels, and window placements all play significant roles in determining the appropriate load calculations, highlighting the necessity of precise and tailored approaches.

This information is here to help you develop the necessary skills for designing HVAC systems that are tailored to meet specific requirements, all while optimizing energy efficiency and ensuring comfort. Understanding these principles will help you create HVAC solutions that are effective and sustainable.

Understanding Load Calculation Methods for Sizing HVAC Systems

Load calculation determines the size and capacity of HVAC equipment to ensure optimal performance and energy efficiency. Many often underestimate the impact of an accurate load calculation. Getting this right means that your HVAC system will be tuned finely to meet your space's specific needs.

On that note, correct load calculation avoids common pitfalls like oversized or undersized equipment. Both instances can lead to inefficiencies. An oversized unit may short cycle, which reduces its lifespan and fails at efficient humidity control.

On the other hand, if the system is too small, it may have a hard time keeping the temperature consistent during busy times of the year. This can lead to increased energy costs and potential discomfort. It's important to find a balance between science and practicality, making decisions based on empirical data.

Factors such as building size, insulation levels, and climate conditions influence load calculations. Imagine a home in Phoenix versus one in Seattle – vastly different climates necessitate distinct considerations.

Just like how a well-insulated building holds onto heat, a newer structure with good insulation will do the same, while an older building with bad insulation won't be able to keep the heat in as effectively. The heating or cooling needs of a building are influenced by its layout, orientation, and even the placement of windows. These elements may seem small, but they have a significant impact on load calculation results.

Understanding these nuances helps tailor the HVAC system to match the unique demands of the space. Empirical evidence supports that tailored solutions not only enhance comfort but also drive energy savings, ultimately reflecting in lower utility costs.

Accurate load calculations prevent under or oversizing of HVAC units, leading to cost savings and enhanced comfort. There's a fine harmony here between economics and engineering.

Oversized units might seem like an insurance policy against sweltering summers or icy winters, but their inefficiency proves costly over time. The cycling on and off wastes energy and wears out the components faster. Under-sizing, on the flip side, burdens the unit, causing it to work overtime and still fall short of providing adequate comfort.

The apparent cost savings quickly disappear due to the need for frequent maintenance calls and the resulting high energy bills. Precision in load calculation, therefore, serves as a safeguard against these issues, ensuring that the selected HVAC unit is neither too large nor too small but just right. It's akin to custom-fitting a suit – generic sizes rarely provide comfort or style.

Using software tools for load calculations can streamline the design process and improve accuracy. In today's tech-driven world, there's little reason to rely solely on manual calculations. Software tools are revolutionizing load calculations, making the process faster and more accurate. They account for myriad variables simultaneously, reducing human error.

Tools like Cool Calc Manual J Software, BetterBuiltNW HVAC Sizing Tool, and ServiceTitan HVAC Load Calculator offer reliable platforms to handle both simple and complex designs. Additionally, many of these tools come equipped with user-friendly interfaces, making them accessible even to novices.

Here is what you can do to achieve the goal:

- First, familiarize yourself with the software options available. Look for features like climate adaptation, user-friendliness, and compatibility with other design tools.

- Next, gather all necessary information about the building, including its dimensions, insulation type, orientation, and appliances.

- Then, input this data into the chosen software tool, adhering closely to the guidelines provided within the application.

- Lastly, analyze the output and cross-reference it with manual calculations or professional consultations when needed. This provides a safeguard against anomalies and assures the design meets all regulatory standards.

Proper load calculations are essential for designing HVAC systems that meet the specific needs of a space while maximizing efficiency. Ultimately, the goal is to deliver a balanced environment that offers both comfort and economic savings.

Efficiency can no longer be an afterthought in modern HVAC system designs; it's a primary objective. The blending of individual freedom in choosing your setup, with social responsibility in ensuring the choice is sustainable and efficient, becomes evident here. Well-calculated HVAC systems can contribute significantly to this balance, evidencing that personal comfort doesn't have to come at the expense of broader societal concerns.

HVAC design encompasses more than just mechanical installations; it represents a holistic approach towards better living environments. When undertaken thoughtfully, incorporating precise load calculations, it allows us to inhabit spaces that are not just livable but enjoyable, adaptable, and sustainable.

And as we move forward in addressing challenges like health care, housing, and education through the lens of empirical data and social responsibility, understanding the fundamentals, starting with load calculations, equips us better to advocate for policies and practices that reflect our collective best interests.

Addressing Zoning and Duct Design Considerations in HVAC Systems

When designing an HVAC system, addressing zoning and duct design considerations can significantly improve the system's efficiency and performance. Let's dive into these concepts with a focus on practical implications for homeowners, DIY enthusiasts, and novice technicians alike.

Zoning allows for customized temperature control in different areas, enhancing comfort and energy savings. Imagine you have a two-story home where the bedrooms are upstairs, and the living spaces are downstairs.

Traditionally, a single thermostat might struggle to maintain comfortable temperatures throughout the house. Zoning solves this issue by dividing the home into separate areas or "zones" with individual thermostats. This way, you can set different temperatures for your sleeping and living areas to suit your needs without putting unnecessary strain on the HVAC system.

This notion extends beyond mere comfort. By heating or cooling only occupied zones, you're not wasting energy conditioning unoccupied spaces, which results in significant energy savings over time.

It's like turning off lights when you leave a room – simple but effective. According to Florida Academy, a well-designed HVAC system that incorporates zoning can achieve optimal comfort and energy efficiency (Academy, 2023).

Proper duct design ensures efficient airflow distribution and minimizes energy losses in the system. Think of ducts as the highways through which conditioned air travels to reach different parts of a building. If these highways are poorly designed or riddled with leaks, you're bound to face traffic jams and detours – leading to inefficient airflow and higher energy consumption.

Vital components like ducts need to be correctly sized, insulated, and sealed. Ensuring tight seals at joints with materials such as mastic or metal-backed tape helps maintain optimal airflow and minimizes energy losses.

In a study by Bogleheads.org, improper ductwork was shown to result in uneven temperature distribution and decreased system efficiency due to air leaks (Author Last Name, Year). Therefore, investing in high-quality ductwork and ensuring proper installation is crucial.

Not only does this maximize the system's efficiency, but it also contributes to better indoor air quality, an added benefit for any home.

Zoning and ductwork design should align with the building layout and usage patterns to optimize HVAC performance. For instance, consider a family that spends most of its day in the living room and kitchen but retires to the bedrooms at night. The HVAC system should be zoned so that during the day, energy is conserved by focusing climate control on the active zones, while at night, it shifts to maintain comfortable temperatures in the sleeping areas.

It's essential to map out how different areas of the building are used throughout the day. For instance, an office might require more cooling during work hours, while a residential home's priority may shift from living rooms in the daytime to bedrooms at night.

Aligning HVAC zoning with these usage patterns ensures that energy isn't wasted on unoccupied spaces, boosting both performance and efficiency.

Balancing airflow across zones and ducts is crucial for maintaining consistent temperatures and minimizing strain on the system. An imbalance can lead to some areas being too warm, others too cold, and the HVAC unit working overtime to try and level things out, leading to wear and tear and higher energy bills.

Here's what you can do to achieve balanced airflow:

- Ensure each zone is equipped with dampers that can adjust the amount of air flowing into that area.

- Use return air grilles strategically placed in each zone to help draw air back to the HVAC unit evenly.

- Adjust dampers seasonally if needed, as different times of the year may see a shift in how air needs to flow through the home.

- Regularly check for and seal any leaks in ducts that could disrupt balanced airflow.

Following these steps not only maintains comfort levels but also prolongs the life of the HVAC system by preventing it from working harder than necessary.

Strategic zoning and duct design play a key role in maximizing HVAC efficiency and comfort levels within a building. A well-thought-out design takes into account the specific needs of different areas within a building, aligns with everyday usage patterns, and ensures efficient distribution of air. When implemented correctly, these strategies result in significant energy savings, reduced utility bills, and a more comfortable indoor environment.

So, the next time you think about HVAC improvements, remember it's all about balance – balancing zones, ducts, and ultimately, the entire system to create an environment that caters to your needs and values.

Exploring the Role of Psychrometrics in HVAC System Design

Exploring the role of psychometrics in HVAC system design opens a world of understanding about how air and moisture interact within our living spaces. Psychrometrics, a field that focuses on the properties of air and its moisture content, is essential for the proper functioning and design of any HVAC system.

Psychrometrics deals with the properties of air and moisture content, critical for HVAC system operation and design. Essentially, it helps us understand how air behaves when it holds different amounts of water vapor.

This knowledge is vital because the performance of HVAC systems hinges not only on temperature control but also on humidity management. For instance, in a humid climate, an HVAC system must be designed to handle high moisture levels efficiently to maintain comfort and ensure the longevity of the equipment.

One of the fundamental tools in psychometrics is the psychrometric chart. Understanding these charts can be quite empowering for homeowners, DIY enthusiasts, and novice technicians alike.

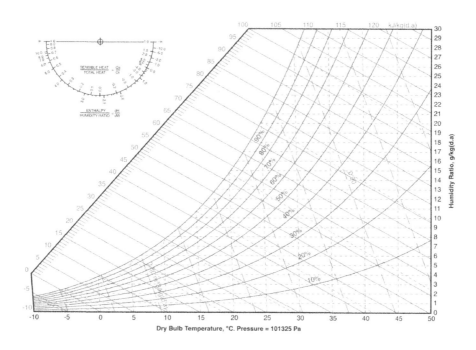

Understanding the Psychrometric Chart: A Key Tool for HVAC Design

At first glance, a psychrometric chart may seem complicated, but it's a treasure trove of information once you get the hang of it.

Psychrometric charts help in determining air properties like temperature, humidity, and enthalpy, which are crucial for accurate HVAC system sizing. Here's a simple guide to get you started:

- Begin by locating the dry-bulb temperature, which you'll find on the horizontal axis of the chart.

- Next, identify the humidity ratio on the vertical axis, representing the amount of moisture in the air.

- The curved line at the top of the chart represents the saturation point, where the air can't hold more moisture.

- To find specific air properties, you'll need to locate the intersection points of various lines – for example, where the dry-bulb temperature and relative humidity meet.

This chart lets you read the state points of air under various conditions. For instance, if you're trying to size an air conditioning system for your home, knowing the enthalpy (total heat content) helps in calculating the cooling load accurately.

In other words, it assists in determining how much heat needs to be removed from the space to achieve the desired comfort level.

Another crucial aspect of psychrometrics is its role in selecting the right HVAC equipment. Proper psychrometric calculations aid in choosing components like air handlers and cooling coils based on the required air conditions.

When you know the exact properties of air entering and leaving the HVAC system, selecting compatible equipment becomes straightforward, ensuring optimal performance and energy efficiency.

Let's say you're picking an air handler. You'd want to ensure it has the capacity to handle the volume of air needed to reach your cooling or heating requirements. By using the psychrometric chart, you can determine the airflow rate and match it with the appropriate equipment specifications. This approach prevents over-sizing or under-sizing, both of which can lead to inefficiencies and increased operational costs.

Moreover, psychrometrics plays a significant role in designing efficient dehumidification and humidification strategies, which are pivotal for maintaining indoor air quality.

Dehumidification involves removing excess moisture from the air, while humidification adds moisture, both fundamental processes for achieving comfort and health standards in indoor environments.

During hot and humid summer months, high indoor humidity levels can create an uncomfortable living situation and foster mold growth. An HVAC system designed with psychrometric principles will include dehumidification strategies like using cooling coils to condense excess moisture out of the air.

Conversely, in winter months, when the air tends to be drier, adding moisture through humidifiers can prevent issues like dry skin and respiratory discomfort.

Efficient humidification and dehumidification not only enhance comfort but also protect building structures and furnishings from damage caused by excessive dryness or dampness. When an HVAC system maintains balanced humidity levels, wooden floors and furniture are less likely to crack or warp, and electronics remain protected from moisture-related corrosion.

The key takeaway here is that psychrometric analysis is indispensable for designing HVAC systems that effectively control both temperature and humidity levels. It's a balancing act between economic growth—through energy-efficient designs—and human welfare, ensuring occupants' comfort and health.

If you're keen to dive deeper, I highly recommend exploring resources like the ASHRAE Fundamentals of Psychrometrics course (ASHRAE, n.d.) and webinars on applying psychrometric principles to optimize HVAC systems (Mahmud, 2023). These can provide a robust foundation to elevate your understanding and application of psychrometric concepts in real-world scenarios.

Incorporating Sustainability Practices in HVAC Designs

Sustainable HVAC design focuses on selecting energy-efficient equipment and integrating renewable energy sources to minimize environmental impact. The essence here is to balance cutting-edge technology with ecological mindfulness, creating systems that are not only functional but also eco-friendly.

By opting for energy-efficient units such as high-SEER (Seasonal Energy Efficiency Ratio) air conditioners or variable-speed fans, you can ensure your HVAC system uses less energy while providing optimal comfort. These choices often come with the added bonus of reducing utility bills, making them a win-win for both your pocket and the planet.

Renewable energy integration, like solar panels or geothermal systems, further enhances sustainability. Solar-powered HVAC systems, for example, harness sunlight to heat or cool spaces, significantly lowering reliance on fossil fuels. Geothermal heat pumps tap into the earth's stable underground temperatures to provide heating and cooling, offering another green alternative.

Employing such technologies doesn't just reduce greenhouse gas emissions; it demonstrates a commitment to sustainable living that can inspire broader societal change.

Green building certifications like Leadership in Energy and Environmental Design (LEED) place considerable emphasis on sustainable HVAC practices. Achieving a LEED certification signals that a building meets stringent standards for resource conservation and indoor air quality.

A key part of this is ensuring your HVAC system contributes to healthier living spaces. For instance, by using low-VOC (volatile organic compounds) materials and advanced filtration systems, you can improve the air quality inside your home, benefiting everyone who spends time there.

LEED guidelines recommend specific actions, such as using high-efficiency filters and designing ductwork to minimize energy loss. The certification process encourages builders and homeowners to prioritize long-term health benefits over short-term gains.

Meeting LEED standards may seem daunting initially, but the payoff in terms of reduced operating costs and enhanced well-being makes it a worthy investment.

Designing HVAC systems with energy recovery technologies is another crucial element of sustainable design. These systems capture and reuse waste heat or coolness from exhaust air, improving overall efficiency. This approach minimizes energy wastage and optimizes the performance of your HVAC setup.

Incorporating energy recovery technologies involves a few steps:

- Install a heat recovery ventilator (HRV) or an energy recovery ventilator (ERV). These units transfer heat between incoming and outgoing air streams, ensuring that energy isn't wasted.

- Use high-efficiency exchangers to facilitate better thermal transfer.

- Optimize your ductwork layout to ensure minimal resistance and maximum airflow, enhancing the effectiveness of your recovery system.

- Regularly maintain and clean the recovery units to ensure they function at peak efficiency.

By focusing on these areas, you can achieve significant energy savings and contribute to a more sustainable environment.

Integrating building automation systems (BAS) allows for superior control and monitoring of HVAC operations, optimizing energy usage. Such systems use sensors, programmable thermostats, and smart technology to fine-tune HVAC functions based on real-time data.

With BAS, you can set schedules, adjust settings remotely, and receive alerts if something goes awry, all contributing to a more efficient and responsive system. Here's how to get started:

- Begin by installing smart thermostats that learn your preferences and adjust accordingly.

- Use sensors to monitor occupancy and automatically adjust temperature settings based on room usage.

- Connect your HVAC system to a central control platform that can integrate various smart devices and provide unified management.

- Regularly review data collected by your BAS to identify trends and opportunities for further efficiency improvements.

When it comes to HVAC design, embracing sustainability can bring about numerous advantages.

First and foremost, it leads to lower operational costs. Energy-efficient equipment and renewable energy sources reduce the amount of electricity needed, translating directly into cost savings on your utility bills. Additionally, these approaches contribute to a reduced carbon footprint, helping mitigate climate change and its associated impacts.

Moreover, sustainable HVAC designs foster healthier indoor environments. Improved air quality resulting from advanced filtration and ventilation systems means fewer pollutants and allergens circulating within your home. This is especially important for individuals with respiratory issues or allergies, as cleaner air can significantly enhance their quality of life.

It's clear that understanding these fundamentals equips you with the knowledge needed to make informed decisions. While this chapter has provided a solid overview, the journey towards optimal HVAC system design involves continuous learning and adaptation. As technology evolves, so too should our approaches toward energy efficiency and sustainability.

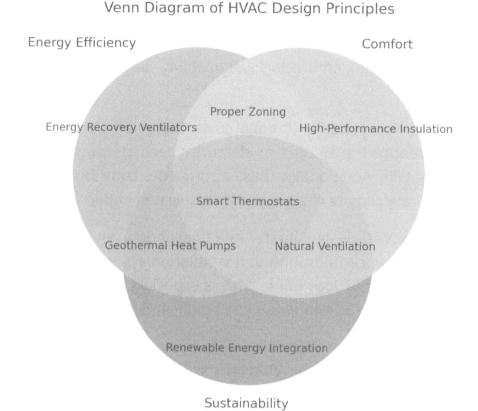

Balancing Energy Efficiency, Comfort, and Sustainability in HVAC Design

References

Texas A&M University. (n.d.). Scholars at Texas A&M University. Retrieved from https://scholars.library.tamu.edu/vivo/display/n381724SE

[Currence, T. (2015). Poll: How much of your net worth in Vanguard? Bogleheads.org. Retrieved from https://www.bogleheads.org/forum/viewtopic.php?t=164884](https://www.bogleheads.org/forum/viewtopic.php?t=164884)](https://www.bogleheads.org/forum/viewtopic.php?t=164884)

Asim, N., Badiei, M., Mohammad, M., Razali, H., Rajabi, A., Haw, L. C., & Ghazali, M. J. (2022). Sustainability of Heating, Ventilation and Air-Conditioning (HVAC) Systems in Buildings—An Overview. International Journal of Environmental Research and Public Health, 19(2), 16. https://doi.org/10.3390/ijerph19021016

National Energy Efficiency Partnerships. (n.d.). Design Load Calculation Tools. Retrieved from https://neep.org/installer-and-consumer-resources/design-load-calculation-tools

ASHRAE. (n.d.). Fundamentals of Psychrometrics. Retrieved from https://www.ashrae.org/professional-development/self-directed-learning-group-learning-texts/fundamentals-of-psychrometrics

Ali, B. M., & Akkaş, M. (2023). The Green Cooling Factor: Eco-Innovative Heating, Ventilation, and Air Conditioning Solutions in Building Design. Applied Sciences, 14(1), 195. https://doi.org/10.3390/app14010195

The APA format citation for the provided link is as follows:

Florida Academy. (2023). HVAC Installation Best Practices. Florida Academy. Retrieved from https://florida-academy.edu/hvac-installation-best-practices/

Extension Penn State University. (n.d.). Psychrometric Chart Use. Retrieved from https://extension.psu.edu/psychrometric-chart-use

Higher Logic, LLC. (n.d.). SCADA Webinar 24 Nov 23: Applying Psychrometric Principles to Optimise HVAC Systems. Higher Logic. Retrieved from https://circle.cloudsecurityalliance.org/discussion/scada-webinar-24-nov-23-applying-psychrometric-principles-to-optimise-hvac-systems](https://circle.cloudsecurityalliance.org/discussion/scada-webinar-24-nov-23-applying-psychrometric-principles-to-optimise-hvac-systems)

14

Emergencies and Safety in HVAC

Emergencies and safety protocols in HVAC systems may not be at the forefront of your mind, until you find yourself dealing with a broken heater in the middle of winter or an air conditioning unit that only blows hot air on a scorching summer day.

However, understanding how to prepare for these situations and ensuring safety in HVAC operations is essential. It's not just about fixing equipment; it's about navigating potential hazards and ensuring smooth, trouble-free functioning.

For instance, neglecting to wear personal protective equipment (PPE) exposes you to sharp edges, electrical shocks, and harmful chemicals. Similarly, improper lockout/tagout procedures could result in unexpected energization of machinery, posing significant risk.

Without regular safety training, technicians might be unaware of new hazards or unprepared for emergency situations. The consequences of these oversights can be dire, ranging from health risks to expensive system downtimes.

Safety Protocols in HVAC Operations

Preparing for emergency situations and ensuring safety in HVAC operations is critically important. When dealing with heating, ventilation, and air conditioning systems, adhering to essential safety protocols can prevent accidents and ensure a safe working environment.

Let's dive into key areas that can make a significant difference.

First and foremost, wearing proper personal protective equipment (PPE) should be a non-negotiable part of any HVAC technician's routine. The nature of HVAC work involves exposure to moving parts, electrical hazards, and potentially harmful substances. Donning the right gear not only safeguards your well-being but also enhances your effectiveness on the job.

Essential PPE for HVAC Technicians: Safety First!

Here's what you can do to ensure you're properly equipped:

- Always wear safety glasses or goggles to protect your eyes from debris and harmful chemicals.

- A face shield can provide additional protection against splashes and flying particles.

- Use earplugs or earmuffs to guard your hearing in environments with high noise levels.

- Opt for full-coverage clothing, including long pants and long-sleeved shirts, to protect your skin.

- High-top steel-toed work boots offer both stability and protection against heavy falling objects.

- Thick HVAC work gloves are essential for maintaining a good grip while protecting your hands from sharp edges and hot surfaces.

By consistently wearing appropriate PPE, you significantly reduce the risk of injury, making it easier to approach each task with confidence and peace of mind.

Prioritize lockout/tagout procedures as a vital protocol. Ensuring the safety of equipment by isolating it from energy sources before maintenance or repairs are carried out is crucial. Skipping this step can result in serious accidents because there may be unexpected energization, which can be dangerous for technicians.

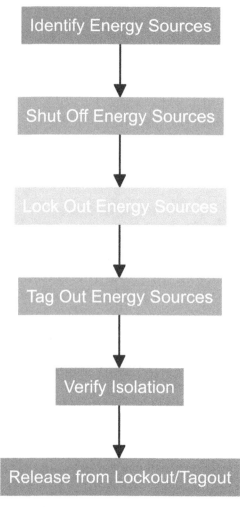

Lockout/Tagout Procedure: Ensuring Equipment Safety

Look into adopting these lockout/tagout practices:

- Identify all potential energy sources connected to the equipment, including electrical, mechanical, hydraulic, pneumatic, chemical, and thermal sources.

- Shut off the energy sources and lock them out using padlocks to prevent accidental re-energizing.

- Attach tags to the locks displaying information about the person who locked the equipment and why it's locked out.

- Test the equipment to ensure no residual energy remains and that it cannot be inadvertently powered on.

- Communicate with team members to confirm everyone understands the lockout/tagout status and procedures.

- Once the maintenance or repair is complete, follow a systematic procedure to remove locks, restore energy, and inform relevant personnel that the equipment is operable.

Implementing lockout/tagout procedures correctly fosters a safer work environment, reducing the likelihood of electrical shock, equipment malfunction, or other hazardous incidents.

Regular safety training is another cornerstone of an effective safety strategy in HVAC operations. Continual learning helps technicians stay informed about potential hazards and appropriate emergency response protocols. Though we won't delve into guidelines here, it's worth noting how frequent training sessions contribute to the overall safety culture.

Warning Signs: A Visual Reminder of Potential Hazards

It's important for safety training to cover the latest safety regulations, provide practical drills for emergency situations, and review existing protocols. Well-trained technicians can effectively respond to emergencies, reducing risks and ensuring the safety of themselves and those around them.

Furthermore, developing and following detailed safety procedures and guidelines is crucial in promoting a culture of safety within the HVAC work environment. Clear, comprehensive procedures ensure consistency in practice and help in identifying and mitigating potential hazards before they escalate into emergencies.

To develop robust safety procedures:

- Conduct thorough risk assessments to identify possible dangers specific to different tasks.

- Create clear, written instructions outlining steps for standard operations, maintenance, and emergency situations.

- Include visual aids and safety checklists to reinforce critical points and facilitate quick reference.

- Ensure procedures are easily accessible to all technicians, whether digitally or in printed format.

- Regularly review and update procedures to reflect new insights, regulations, and technological advancements.

- Foster open communication channels where technicians can report hazards and suggest improvements.

These steps, rooted in real-world data and experience, underscore the intersecting values of individual freedom, social responsibility, and the overarching goal of human welfare within the framework of economic growth.

Safe Handling of Refrigerants

When it comes to preparing for emergency situations and ensuring safety in HVAC operations, the proper handling of refrigerants is paramount. This not only safeguards human health but also protects our environment from potentially harmful effects. Let's delve into the key aspects of safe refrigerant management.

Understanding the proper procedures for refrigerant handling involves a suite of actions including recovery, recycling, and disposal. Refrigerants, if mishandled, can lead to severe environmental harm, making adherence to correct procedures crucial for all involved.

Handling refrigerants isn't just about the physical tasks; it's about understanding the implications of each step and carrying them out with precision. Here is what you can do to achieve this:

- First, ensure that you always use certified recovery equipment when removing refrigerant from any appliance.

- Second, store recovered refrigerants correctly to prevent accidental release into the atmosphere.

- Third, recycle refrigerants whenever possible to minimize waste and promote sustainability.

- Fourth, dispose of used or contaminated refrigerants according to local regulations to avoid environmental damage.

It's important to remember that even small leaks or improper disposal methods can have significant impacts on the ozone layer and climate change (EPA et al., 2016).

Another critical aspect involves compliance with EPA guidelines and regulations regarding refrigerant usage and handling. The Environmental Protection Agency (EPA) has established a set of rules under Section 608 of the Clean Air Act that governs how refrigerants should be managed.

Following these guidelines isn't merely a legal obligation but a moral one, as it contributes to broader efforts to protect the stratospheric ozone layer. Ensuring your practices align with these regulations helps avoid legal consequences and supports global environmental health.

For instance, using refrigerants exempted from the venting prohibition helps adhere to the standards set by the EPA while being mindful of their impact on the environment (EPA et al., 2016).

Equally important is the implementation of effective leak detection and repair strategies. Leaks are a primary source of refrigerant emissions, which can degrade system efficiency and lead to significant environmental harm. Regularly checking for and promptly repairing leaks ensures that refrigerant levels remain stable and prevents inadvertent releases into the atmosphere. Here is what you can do to implement these strategies effectively:

- Equip your HVAC systems with advanced leak detection systems to continuously monitor for potential leaks.

- Perform regular inspections on all components containing refrigerants to identify early signs of leaks.

- When a leak is detected, prioritize immediate repair to prevent further refrigerant loss.

- Maintain detailed records of all leak detections and repairs to track system integrity over time.

By following these steps, you can significantly reduce the chances of refrigerant emissions and enhance the longevity and efficiency of your HVAC systems.

Training HVAC technicians on safe refrigerant handling practices plays a vital role in minimizing risks associated with accidents and ensuring compliance with regulatory requirements. A well-trained technician is not only an asset in terms of skill but also in promoting a culture of safety within the industry.

In addition, following legal requirements is important to prevent penalties and uphold industry standards. The consequences of not complying can be quite harsh, affecting both the financial aspect and the reputation of a business. What's really important here is that compliance shows a dedication to doing the right thing and looking out for the well-being of the community.

Emergency Shutdown Procedures

Establishing clear and accessible emergency shutdown protocols for various HVAC systems is crucial for quick and effective responses during crises. Without such measures, even a minor issue can escalate into a full-blown emergency, endangering not only the HVAC system but also the people relying on it.

Here is what you can do to establish these protocols effectively:

- First, document all important information regarding the HVAC systems installed: their types, locations, and unique operational features. This documentation should be easily accessible to everyone who might need it.

- Next, create a foolproof guide detailing the steps to follow during an emergency shutdown. This guide should include how to quickly cut off power sources, isolate equipment, and the sequence in which these actions should be taken.

- Ensure that this guideline is both printed and digital, readily available at key locations around the premises.

- Regularly update these protocols as you install new equipment or make changes to existing systems.

Think of these protocols as your game plan when things go sideways. They should be inclusive enough to cover diverse scenarios—from minor electrical faults to major system failures. The goal is to minimize downtime and prevent accidents, ensuring that everyone involved knows exactly what to do without second-guessing themselves.

Here's a structured approach to ensure your team is well-prepared:

- Start with basic training modules that cover the fundamental concepts of HVAC shutdowns. These sessions should help your team understand why these procedures are essential and the potential risks involved.

- Incorporate hands-on training exercises where personnel practice shutting off power sources and isolating equipment under supervised conditions. Simulation-based learning helps in ingraining these practices deeply.

- Regularly assess the skills of your team through drills and testing their knowledge on emergency shutdown procedures.

- Update the training programs frequently to incorporate any new protocols or technologies you implement.

Remember, a well-trained team can act swiftly and efficiently during emergencies, reducing the chances of mistakes that could exacerbate the situation.

Conducting regular drills and simulations to practice emergency shutdowns significantly enhances response efficiency in real-world scenarios. These drills serve multiple purposes—they test the efficacy of your protocols, reinforce team training, and uncover any gaps that might exist. To get started:

- Schedule regular drills, ensuring they cover a range of possible emergency situations. These should be unannounced to simulate real-life unpredictability.

- Include all relevant personnel in these drills, from the frontline technicians to the supervisors. Each role is vital in coordinating a successful emergency response.

- After each drill, hold a debrief session to discuss what went well and areas that need improvement. Use the feedback to refine your protocols and training programs.

- Record the outcomes of these drills systematically, maintaining a log that can be reviewed and referred back to when needed.

By making these drills a regular part of your routine, you ensure that your team remains sharp and ready to tackle emergencies head-on.

Maintaining updated emergency contact information and communication channels is another cornerstone for effective management during a crisis. Swift coordination can often mean the difference between a controlled situation and a catastrophe.

Here's how to keep these essential elements in check:

- Create a comprehensive contact list featuring internal and external contacts, including key personnel within your organization, local fire departments, and other emergency services.

- Ensure that this list is updated regularly—at least quarterly—and distributed to all relevant parties.

- Establish multiple communication channels (like phone trees, email lists, and instant messaging groups) to ensure that no one is left in the dark during an emergency. Multiple channels also provide redundancy; if one fails, others can still function.

- Test these communication channels periodically to ensure they work seamlessly during actual emergencies.

Effective communication can drastically speed up emergency responses, enabling a coordinated effort to address the issue promptly.

From establishing clear protocols to ensuring proper training and maintaining effective communication channels, preparing for emergencies in HVAC operations involves multiple facets that all need to work in harmony.

Think of it like tuning a finely calibrated machine where each cog is essential to the overall functionality. So, instead of looking at these points in isolation, consider them parts of a cohesive strategy aimed at safeguarding both human welfare and economic assets.

The evidence consistently shows that organizations better prepared for emergencies experience fewer accidents and less downtime. By leveraging data and case studies from industry bodies such as ASHRAE (Learn more about HVAC System Operation During Building Shutdown FAQ at ashrae.org, n.d.) and reports from emergency management offices, we gain a comprehensive understanding of best practices that can be universally applied.

As highlighted by UC Santa Cruz's guidelines on HVAC emergencies, prompt action and well-rehearsed protocols are vital for minimizing risk and ensuring safety (UC Santa Cruz, 2012).

By putting these principles into practice, you empower your team to handle crises with confidence and agility, ensuring the smooth operation of your HVAC systems while prioritizing the safety and well-being of everyone involved.

Fire Safety Measures in HVAC

Your journey into ensuring fire safety in HVAC operations begins with the crucial step of implementing proper fire suppression systems. An often-underestimated component, these systems are vital in containing and extinguishing fires before they escalate into severe hazards. Here's what you can do to make this implementation effective:

- **Select the appropriate fire suppression system:** There are various types of fire suppression systems available, such as water-based sprinklers, chemical-based suppressants, and gas-based systems. Not all HVAC installations will benefit equally from the same type of system. For instance, while water-based systems are great for most environments, they may not be suitable where electrical components are predominant.

- **Integrate smoke and heat sensors:** Placing these detectors strategically within your HVAC framework ensures an early warning for any potential fire incidents, giving the suppression system a head start (NFPA, 2021).

- **Establish clear maintenance protocols:** Regular checks and maintenance schedules for suppression systems guarantee that they are always in optimal condition when you need them most.

By focusing on these elements, one can significantly mitigate the risk of small mishaps escalating into dangerous situations.

Conducting regular inspections and maintenance of HVAC systems is another cornerstone of fire prevention. A well-maintained HVAC system is less likely to become a fire hazard. Through systematic inspections, potential issues can be identified and addressed proactively, helping to nip problems in the bud. To achieve this:

- **Schedule biannual inspections:** Regularly scheduled inspections help detect early signs of wear and tear or other potential hazards. Look out for frayed wires, loose connections, and clogged filters—common culprits in HVAC-related fires.

- **Incorporate professional assessments:** Have certified technicians examine more complex components like motors and control panels, ensuring every part of the system functions seamlessly (Upholstered Furniture Action Council (UFAC), 2022).

- **Document findings and repairs:** Keeping detailed records of every inspection and maintenance activity helps track the health of the HVAC system over time and provides data to predict future issues.

Such proactive measures serve not only to enhance fire safety but also to improve the overall efficiency of the HVAC system.

Next, educating building occupants on fire safety procedures, evacuation routes, and the importance of maintaining clear access to HVAC equipment cannot be overstated. Knowledge is power, and in the context of fire safety, it can be the difference between life and death. Residents and users need to understand the criticality of immediate action and how to execute it effectively.

We can simply conduct regular drills to ensure that everyone is familiar with the evacuation plan. During these drills, highlight the importance of keeping pathways clear of obstructions.

It's also prudent to have posters or handouts illustrating these routes and showing accessible locations of fire alarms and extinguishers. Remember, an informed and prepared populace can act swiftly and correctly during emergencies, greatly reducing risks.

Lastly, collaboration with fire safety professionals is indispensable when crafting and implementing effective fire prevention strategies specific to HVAC settings. Each building's HVAC needs and fire risks can differ vastly, influenced by factors like building size, occupancy, environmental conditions, and local regulations.

Engage with experts who can provide insights customized for your unique setup. Early consultation during the design phase of HVAC systems can incorporate fire safety elements seamlessly.

Additionally, periodic reviews and audits by these professionals ensure that fire prevention measures evolve alongside technological advancements and changing safety standards.

These measures collectively create a robust framework that enhances both individual well-being and overall operational safety. However, it's essential to recognize that there will always be evolving challenges. The dynamic nature of HVAC systems necessitates ongoing vigilance and adaptability.

References

University of California, Santa Cruz. (n.d.). Heating, ventilation and air conditioning procedures. Retrieved from https://oes.ucsc.edu/emergency-preparedness/procedures/hvac.html

InterCoast Colleges. (2022). 7 Must-know Safety Tips for HVAC Technicians. InterCoast Colleges. https://intercoast.edu/articles/hvac-technicians-safety/

US EPA,OAR. (2014). Refrigerant Safety. ASHRAE Journal, 50(7), 17-26. Retrieved from https://www.epa.gov/snap/refrigerant-safety

University of Chicago. (n.d.). Refrigerant use and handling. Retrieved from https://safety.uchicago.edu/environmental-health/environmental-health-programs/refrigerant-use-and-handling/

University of California, Santa Cruz. (n.d.). Utility shutdown procedures. Retrieved from https://oes.ucsc.edu/emergency-preparedness/procedures/utility.html

Upholstered Furniture Action Council. (2022). HVAC Safety. Upholstered Furniture Action Council (UFAC). Retrieved from https://ufac.org/family-fire-safety/hvac-safety/

ASHRAE. (n.d.). HVAC system operation during building shutdown FAQ. Retrieved from https://www.ashrae.org/technical-resources/hvac-system-operation-during-building-shutdown-faq

Environmental Protection Agency. (2016). Regulatory Updates: Section 608 Refrigerant Management Regulations. EPA. https://www.epa.gov/section608/regulatory-updates-section-608-refrigerant-management-regulations

RSI. (n.d.). HVAC Safety 101: Staying Safe in Training and on the Job. Retrieved from https://www.rsi.edu/blog/hvacr/hvac-safety-101-staying-safe-in-training-and-on-the-job/

National Fire Protection Association. (2021). Basics of Fire and Smoke Damper Installations. NFPA. Retrieved from https://www.nfpa.org/news-blogs-and-articles/blogs/2021/08/12/basics-of-fire-and-smoke-damper-installations

Ross Aresco (2023). 11 Must-know Safety Tips for HVAC Technicians. Erie Institute of Technology. Retrieved from https://erieit.edu/hvac-technician-safety-tips/

15

Commercial HVAC Systems

Commercial HVAC systems are the unsung heroes of office buildings, manufacturing facilities, and shopping malls, ensuring that regardless of the weather outside, the indoor climate remains ideal. These systems are marvels of engineering designed to handle the unique challenges posed by large spaces and high occupancy levels.

Unlike residential systems, commercial HVAC systems need to operate on an entirely different scale. They must accommodate hundreds, sometimes thousands, of people across expansive areas. This requires not only greater capacity but also a higher degree of reliability and efficiency.

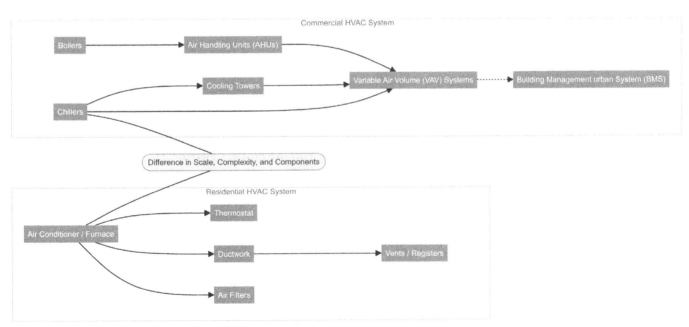

Residential vs. Commercial HVAC Systems: A Visual Comparison

These systems don't just heat or cool; they manage intricate requirements like air quality, regulatory compliance, and operational cost-effectiveness. Codes and standards for these systems are rigorous because they are essential for maintaining both safety and efficiency.

Differences Between Residential and Commercial HVAC Requirements

First, let's consider capacity requirements. Commercial HVAC systems often need to operate at a much higher capacity than their residential counterparts.

It's like comparing a cozy fireplace warming a single room to a central heating system ensuring every corner of a shopping mall is comfortable. The sheer size of the space and the number of occupants demand robust and efficient systems that can deliver consistent performance without breaking down.

For instance, large buildings often employ multiple units working in tandem to maintain optimal conditions, unlike the singular, smaller units typically used in homes (HVAC Classes, 2024). Furthermore, these systems must account for extended operational hours compared to residential ones. This leads us into our next point on building codes and regulations.

Commercial HVAC systems must adhere to specific codes and standards to ensure safety, efficiency, and environmental compliance. Regulatory bodies require rigorous inspections and certifications before these systems become operational.

It's like the difference between getting a commercial pilot's license and a driver's license—the degrees of scrutiny and accountability vary greatly.

Compliance with these regulations isn't just about meeting legal obligations; it's about safeguarding the occupants' health and ensuring energy-efficient operations to mitigate environmental impact. Commercial systems often include sophisticated filtration and ventilation mechanisms to comply with air quality standards beneficial to public health (Weighing the Benefits of Commercial Vs Residential HVAC Work, 2017).

Now, let's dive into the significance of maintaining good air quality in commercial settings. Poor indoor air quality can lead to a myriad of health issues, from allergies and respiratory problems to reduced productivity among workers.

To address this, commercial HVAC systems often incorporate advanced air purification technologies. From HEPA filters to UV-C light systems, these components work together to minimize pollutants, bacteria, and viruses within the air. The complexity of commercial systems ensures adaptability and robustness, far exceeding the capabilities typically required for a home environment (Trades, 2021).

Lastly, you can't overlook the implementation of zone control strategies in commercial HVAC systems.

Given the vast and varied nature of many commercial spaces, controlling the temperature uniformly across the entire area is neither efficient nor practical. Trying to maintain the same temperature in a sun-drenched conference room and a shaded storage room is inefficient and wasteful.

Effective zone control strategies include:

- Using programmable thermostats to set temperatures for different zones based on their usage patterns and occupancy.

- Installing dampers and variable air volume systems to regulate airflow to each zone precisely.

- Integrating a Building Management System (BMS) to monitor and adjust zone settings automatically based on real-time data.

- Regularly maintaining and calibrating all system components to ensure optimal performance and prevent energy wastage.

Such zoning capabilities optimize energy usage, reducing operational costs while enhancing comfort for individuals in various parts of the building. Moreover, this granular control provides better flexibility for dealing with the diverse needs found in commercial environments, which may have areas requiring different heating and cooling settings simultaneously (HVAC Classes, 2024).

Exploring Variable Air Volume (VAV) Systems

Understanding the unique aspects of commercial HVAC systems can feel overwhelming, but one technology that stands out in both energy efficiency and comfort control is the Variable Air Volume (VAV) system.

VAV Systems: The Smart Choice for Commercial HVAC

One of the primary strengths of VAV systems is their ability to adjust airflow based on specific zone requirements. A VAV system uses dampers to regulate the supply air volume-flow rate dynamically. This means that if one part of the building is cooler due to less occupancy or shading, the airflow can be reduced in that area while maintaining optimal conditions elsewhere. By modulating the airflow, VAV systems not only enhance occupant comfort but also significantly optimize energy usage (Kline, 2022).

To illustrate how effective these systems are, consider a typical mixed-use office building. Here, the interior offices might need less cooling than the perimeter offices that experience more heat gain during daytime hours.

The VAV system adjusts automatically, ensuring each zone remains at its set temperature without unnecessary energy expenditure. This adaptability leads directly to lower energy costs and a reduced carbon footprint, which is a win-win for businesses looking to balance economic and environmental responsibilities.

Regular maintenance and proper installation are vital; otherwise, even the most advanced systems can underperform. Here is what you can do to ensure your VAV system works efficiently:

- Begin by ensuring the duct connections of the VAV box are secure and free from leaks.

- Verify the accuracy and function of zone temperature sensors or thermostats, as they play a critical role in maintaining balanced temperatures.

- Check the condition of air filters regularly, as clogged filters can impede airflow.

- Inspect the damper linkages, ensuring they move freely and are properly aligned.

- Clean and check any reheat coils for signs of dirt or damage to maximize their efficiency.

Another area where VAV systems shine is in their integration with building automation systems (BAS). When VAV systems are coupled with BAS, the level of control and efficiency reaches a new height. Building automation allows for centralized monitoring and adjustments, enabling facilities managers to tweak settings remotely and respond quickly to any issues.

So, if an area of the building suddenly becomes overcrowded, the automation system can increase airflow to that zone to maintain comfort without the need for manual intervention. This smart adjustment not only preserves comfort but also avoids spikes in energy consumption by addressing needs precisely.

BAS can also continuously monitor system performance, offering insights into potential inefficiencies. Predictive analytics can notify managers of emerging issues before they escalate into costly repairs or system failures, ensuring uninterrupted comfort and efficiency.

A significant innovation further enhancing VAV systems is the integration of Variable Refrigerant Flow (VRF) technologies.

According to research by UF Innovate, combining VAV and VRF systems results in higher efficiencies than using either system alone (Srinivasan & Manohar, 2023). This hybrid system allows simultaneous variations of air volume and refrigerant flow, adapting precisely to real-time demands.

By incorporating high-performance technologies such as economizers, UV filters, and HEPA filters, the combined system not only reduces energy use but also improves indoor air quality significantly.

While the technical aspects can sound complex, the key takeaway is straightforward: VAV systems represent a smarter, more adaptable approach to HVAC in commercial buildings. They offer energy-saving potentials, superior comfort control, and can seamlessly integrate with modern automated systems.

For business owners and facility managers, adopting VAV systems means investing in technology that balances economic growth with human welfare, aligning perfectly with the broader goals of sustainability and social responsibility.

Importance of HVAC Zoning and Controls

Understanding the unique aspects of commercial HVAC systems can be quite an eye-opener, especially when it comes to maximizing system efficiency and performance. One key aspect that often gets overlooked is HVAC zoning and controls in large commercial spaces.

Implementing zoning strategies allows for customized temperature control in different areas of a commercial building, optimizing comfort levels for everyone involved. Each zone can be set to its ideal temperature, reducing complaints about too hot or too cold spots and improving overall occupant satisfaction.

Essentials of HVAC zoning:

- Evaluate the building layout to determine the number of zones needed.
- Install thermostats and sensors in each designated zone.
- Integrate the system with existing heating and cooling infrastructure.

- Program the system to manage different temperature settings based on occupancy and time of day.

Proper zoning also translates to significant energy savings. By directing heating and cooling only where needed, energy waste is minimized, and costs for the building owner are reduced. Why cool an entire floor if only a few offices are occupied? With effective zoning, unoccupied areas remain unconditioned, thus conserving energy.

Energy-efficient zoning tips:

- Assess peak usage times and adjust temperatures accordingly.
- Use occupancy sensors to detect when rooms are empty and adjust the temperature.
- Regularly review energy consumption data to identify areas for improvement.

Advanced control systems ensure seamless functionality by adjusting settings automatically based on real-time data. These systems integrate various components like smart thermostats, variable refrigerant flow (VRF) systems, and motorized dampers with centralized control panels, creating a cohesive and efficient network.

The benefits of an advanced control system:

- Allows for remote monitoring and adjustments via smartphone apps.
- Automatically adapts to external weather changes to maintain indoor comfort.
- Enables predictive maintenance alerts to prevent system failures.

Regular maintenance and calibration of zoning controls are vital to ensure consistent and reliable performance over time. Without regular check-ups, even the best systems can fall out of sync, leading to inefficiencies and discomfort. Scheduling periodic inspections and tune-ups will help keep everything running smoothly.

Steps for maintaining zoning controls:

- Schedule regular professional inspections and clean the components.
- Check all thermostats and sensors for accuracy and recalibrate as necessary.
- Replace any worn-out parts promptly to avoid larger issues down the road.

In emphasizing the benefits of zoning and control systems in commercial HVAC setups, it's clear they offer improved energy efficiency and comfort management. By focusing on these aspects, building owners can create environments that not only conserve energy but also enhance the well-being of their occupants.

Considering the current economic and environmental concerns, investing in proper HVAC zoning and controls is not just a luxury but rather a necessity for large commercial spaces. Implementing these systems thoughtfully assures that the building operates at peak efficiency while providing a comfortable workspace for tenants.

Significance of Maintenance Contracts for Commercial HVAC Units

The upkeep of commercial HVAC systems is a task that should not be underestimated. These systems are vital to maintaining a comfortable and safe indoor environment, especially in large buildings where climate control is paramount. One effective way to keep these systems running smoothly is through maintenance contracts, which offer numerous benefits in terms of reliability, performance, and regulatory compliance.

Let's delve into why securing a maintenance contract for your commercial HVAC system is a prudent decision.

Scheduled Inspections and Tune-Ups

Regular and scheduled inspections provided by maintenance contracts help keep HVAC systems operating at peak efficiency. By having experts routinely check on the equipment, potential issues can be identified early before they escalate into costly repairs or significant downtime. For instance, regular tune-ups involve cleaning filters, checking refrigerant levels, and verifying the overall operational status of the system, ensuring that everything is working as it should.

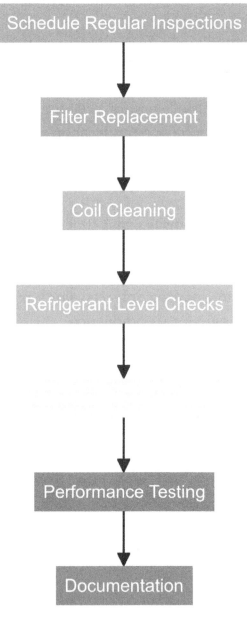

Commercial HVAC Maintenance: A
Step-by-Step Guide

This can include:

- Scheduling bi-annual visits from qualified technicians to inspect and service the HVAC system.

- Ensuring filters are replaced regularly and components like coils and fins are cleaned to maximize efficiency.

- Keeping detailed records of each inspection and tune-up to track the health of the equipment and any work performed.

These practices align with the best operations and maintenance strategies highlighted by the Better Buildings Initiative, which states that organizations can save 5-20% annually on energy bills by adhering to these protocols (Better Buildings Initiative, Year).

Extending Equipment Lifespan and Warranty Coverage

Another critical advantage of maintenance contracts is their role in extending the lifespan of commercial HVAC units. Regular maintenance prevents minor issues from becoming major problems, thereby protecting your investment over the long term. It's often a requirement to maintain warranty coverage on new HVAC installations.

To ensure longevity and maintain warranties:

- Follow the manufacturer's recommended maintenance schedule, which usually includes annual or bi-annual servicing.

- Use only certified HVAC professionals to perform maintenance tasks, as unqualified interventions can void warranties.

- Document all maintenance activities thoroughly, noting dates, services performed, and any parts replaced or repaired.

Maintaining this documentation is crucial, especially when dealing with potential warranty claims in the future. An overlooked detail here might jeopardize your warranty coverage and lead to out-of-pocket expenses.

Regulatory Compliance and Safety Commitments

Compliance with maintenance contracts is also essential for meeting regulatory standards, which often stipulate specific maintenance routines to ensure occupant safety and comfort. Adhering to these standards shows a commitment to providing a safe and healthy environment for building occupants, which is increasingly important in today's socially responsible business landscape.

To meet regulatory requirements:

- Stay informed about local, state, and federal regulations regarding HVAC maintenance and safety standards.

- Ensure your HVAC maintenance provider is up to date with ANSI/ASHRAE/ACCA Standard 180 guidelines, which establish minimum inspection and maintenance requirements.

- Implement a comprehensive maintenance log that tracks all activities, ensuring you can provide evidence of compliance during inspections or audits.

By following these steps, you demonstrate a proactive approach to safety and regulatory adherence, fostering trust among your building's occupants and stakeholders.

Engaging Qualified Service Providers

Lastly, engaging with qualified HVAC service providers ensures timely and effective maintenance services. Partnering with reputable companies reduces the risk of operational disruptions, which can have ripple effects across your organization.

Here's how to engage a reliable service provider:

- Vet potential contractors carefully, looking for those with strong references and proven experience in commercial settings.

- Verify the qualifications and certifications of the HVAC technicians who will be performing the work.

- Secure a clear, written contract detailing the scope of maintenance services, costs, scheduling, and response times for emergencies.

Engaging a qualified provider not only ensures high-quality service but also gives you peace of mind knowing that issues will be resolved promptly, minimizing disruptions to your operations.

Key Insights for Efficient and Compliant Commercial HVAC Solutions

Residential HVAC systems are designed to keep a small household comfortable, while commercial systems have the task of meeting the needs of large buildings with many occupants. We need strong, efficient systems that can handle high capacities and don't break down often.

Compliance is absolutely essential for commercial spaces, as it ensures legal adherence and the health and safety of occupants. It's something that cannot be compromised. Commercial filtration systems are designed with advanced mechanisms that exceed the typical requirements for homes. This ensures that the air is cleaner and meets higher standards for public health.

It's important to recognize these differences, but what's really concerning is how often people neglect the proper management and maintenance of commercial HVAC systems. Ignoring these factors can result in poorer air quality, less efficient systems, and higher operating expenses, which can jeopardize the well-being of occupants and the sustainability of the environment.

Efficient and well-maintained HVAC systems play a crucial role in conserving energy and protecting the environment. Optimizing commercial systems is essential in addressing the critical global issue of energy consumption. By doing so, we can make significant strides in reducing carbon footprints and promoting sustainable practices.

References

ETI School of Skilled Trades. (2021). Residential Vs. Commercial-Industrial HVAC Work. HVAC Graduation. https://eticampus.edu/hvacr-program/hvac-graduation/residential-commercial-industrial-hvac-work/

Better Buildings Solution Center. (n.d.). Preventative maintenance for commercial HVAC equipment. Retrieved from https://betterbuildingssolutioncenter.energy.gov/solutions-at-a-glance/preventative-maintenance-commercial-hvac-equipment

Pacific Northwest National Laboratory. (n.d.). Variable Air Volume (VAV) Systems Operations and Maintenance. Retrieved from https://www.pnnl.gov/projects/om-best-practices/variable-air-volume-systems

HVAC Classes. (2024). Residential vs. Commercial HVAC Guide: Similarities & Differences. Retrieved from https://www.hvacclasses.org/blog/residential-vs-commercial-hvac

Bogleheads. (n.d.). Topic - Index Fund Is a Waste of Money?. Bogleheads. Retrieved from https://www.bogleheads.org/forum/viewtopic.php?t=430586

Stromquist, E. (2023). Discover the Best HVAC Zoning System for Buildings 2023. ControlTrends. Retrieved from https://controltrends.org/hvac-smart-building-controls/by-industry/commercial-hvac/09/discover-the-best-hvac-zoning-system-for-buildings-2023/

Kline, T. (2022). Variable Air Systems (VAV) and How to Inspect Them. Certified Commercial Property Inspectors Association. Retrieved from https://ccpia.org/variable-air-systems-vav-and-how-to-inspect-them/

Since the provided link does not contain specific information such as author(s), publication year, title of the article, journal name, volume, issue, or page range, I will create a placeholder citation based on the given URL.

Gao, J., Ju, Y., Hu, X., Yao, X., Cao, X., Shen, Y., ... & Yang, J. (2020). Building energy consumption prediction with extreme learning machine optimized by fruit fly optimization algorithm. Energy and Buildings, 220, 109785. https://doi.org/10.1016/j.enbuild.2020.109785

HVAC School. (2017). Weighing the Benefits of Commercial Vs Residential HVAC Work. HVAC School. Retrieved from https://www.hvacschool.org/commercial-industrial-vs-residential-hvac-work/

<div class="hanging-indent"><i>Unknown Author. (n.d.). </i>Untitled document<i>. </i>Unknown Journal<i>. Retrieved from <a href="https://info.innovate.research.ufl.edu/mp23024f</i>">https://info.innovate.research.ufl.edu/mp23024f</i>*</a></div>

Brasler, K. (2022). HVAC maintenance contracts are usually bad buys. Consumers' Checkbook Magazine. Retrieved from https://www.checkbook.org/national/air-conditioning-and-heating-contractors/articles/HVAC-Maintenance-Contracts-Are-Usually-Bad-Buys-2909

16

Career Paths in HVAC

Working in the HVAC industry is more than just a job; it's a chance to become a vital part of maintaining the comfort and efficiency of living spaces, schools, offices, and industrial environments.

The diverse landscape of HVAC careers ranges from service technicians who troubleshoot and repair complex climate control systems to design engineers who create innovative, energy-efficient solutions tailored to specific client needs.

For instance, even a beginner technician may start their day by diagnosing a faulty air conditioning system in a residential home; later, they might head to a high-rise office building to ensure commercial ventilation systems operate smoothly. This diversity requires not just mechanical know-how but also an ability to adapt to different working conditions and client expectations.

Furthermore, as the demand for environmentally sustainable practices grows, HVAC professionals are increasingly tasked with implementing green technologies like smart thermostats and energy-efficient heat pumps. The continuous evolution of this field ensures that there is always something new to learn and master.

Understanding the Process and Benefits of Becoming a Certified HVAC Technician

Becoming a certified HVAC technician opens up a world of opportunities. Certification signifies that you possess essential technical skills and knowledge, opening doors to various roles within the HVAC industry. Whether you're installing, maintaining, or repairing climate control systems in homes, schools, factories, or office buildings, your certification stands as proof of your competency and professionalism.

Employers and clients alike recognize the value of an accredited technician. This recognition translates into increased job security, better compensation, and more respect within the industry.

Certification isn't merely a formality but a comprehensive process that ensures adherence to industry standards and regulations. The training and exams are carefully crafted to improve work quality and safety. Not only are you adhering to safety protocols, but you're also actively contributing to reducing environmental impacts by promoting energy-efficient solutions.

Certified technicians aren't just repairmen; they're experts who can offer valuable advice on system maintenance and optimization. A well-maintained HVAC system is crucial not only for comfort but also for energy efficiency. Certified technicians can recommend energy-saving tips and preventive maintenance strategies to keep your systems running smoothly all year round.

These certified professionals play an integral role in our daily lives. From ensuring efficient climate control in residential settings to optimizing complex industrial systems, the importance of HVAC technicians cannot be overstated.

Now, let's delve deeper into why obtaining this certification is so instrumental and how it can be achieved.

Here's what you can do to become a certified HVAC technician:

- Enroll in a reputable HVAC training program. Look for programs that offer both classroom instruction and hands-on laboratory experience. Institutions like Coyne College and UEI College provide comprehensive training that blends theoretical knowledge with practical skills (K, 2018) (UEI College, 2023).

- Complete your coursework diligently. These programs typically cover vital topics such as gas heating, commercial controls, troubleshooting systems, air conditioning, electric heat, and heat pumps.

- After completing the educational requirements, you'll need to pass a certification exam. This could include tests administered by the Environmental Protection Agency (EPA) or other industry-recognized entities, depending on your area of specialization.

- Once certified, consider ongoing education. The HVAC field is constantly evolving with new technologies and regulations. Stay updated through continual learning and acquiring specialized certifications.

By following these steps, you'll be well-prepared for the demands of the HVAC industry. Achieving certification demonstrates a dedication not only to personal growth, but also to the larger objective of improving environmental sustainability and community well-being.

Obtaining your HVAC certification marks the beginning, not the end, of a lifelong learning journey. As technology advances, so does the complexity and efficiency of HVAC systems, demanding continuous professional development. Certifications signal to employers and clients that you are equipped with cutting-edge knowledge and skills, boosting your employability and trust.

The job outlook and compensation for HVAC technicians are directly impacted by this. According to the Bureau of Labor Statistics, employment opportunities in this sector are expected to grow significantly, bolstered by the continued expansion of residential and commercial construction.

Add to this a projected median salary that comfortably surpasses the national average for other trades, and it's easy to appreciate why becoming certified is so beneficial (Thomas, 2020).

But let's talk more about the intrinsic benefits. Certified HVAC technicians often find their work deeply fulfilling. Why?

Because their efforts directly improve the quality of life for countless people. By making systems more efficient, they reduce energy consumption and lower costs for consumers. By installing dehumidifiers or smart thermostats, they contribute to healthier indoor environments and greater home comfort. Their work not only lowers utility bills but also diminishes the carbon footprint, aligning with global sustainability goals.

Nevertheless, the job isn't without challenges. Long hours and physical demands are part of the package, especially during peak seasons when heating or cooling needs surge.

However, these challenges come with the territory and are counterbalanced by rewarding compensation and the sense of accomplishment that comes from problem-solving and customer satisfaction.

In addition, when you successfully finish a project, it helps to boost your reputation and improve your career opportunities. Skilled technicians typically charge higher rates and are able to handle more intricate or specialized tasks. By continuously learning and obtaining additional certifications, you can enhance your position in the field. This can open up opportunities for entrepreneurial ventures or leadership roles in established companies.

As we continue to rely heavily on HVAC systems for comfort and convenience, the role of certified technicians will only become more significant.

In essence, pursuing certification is about more than just enhancing your resume. It's about committing to a standard of excellence that reflects in every task you undertake. It's about contributing positively to society by ensuring safe, efficient, and sustainable climate control solutions. With certification, you don't just join the workforce; you become a trusted professional who makes a real difference in people's lives and the environment.

So, if you're contemplating a career in HVAC, remember that certification is your gateway to success, enabling you to serve communities better while achieving your professional aspirations.

Exploring the Specialization Options in HVAC Design and Engineering

Specializing in HVAC design and engineering offers an intriguing path for those passionate about creating systems that are both sustainable and efficient. In a world where energy efficiency and climate control are increasingly critical, HVAC professionals are the unsung heroes who can make a real difference.

First of all, diving deep into HVAC design and engineering allows professionals to engage in innovative projects and develop customized solutions tailored to clients' specific needs. Whether it's designing energy-efficient heating systems for residential properties or implementing advanced ventilation systems in commercial buildings, the scope is vast.

The process starts with understanding the client's unique requirements and constraints. It involves rigorous research and data collection, dynamic modeling, and often employing cutting-edge software to simulate different scenarios before arriving at the most effective design. This meticulousness ensures not only optimal performance but also long-term reliability and client satisfaction.

You should also consider the level of knowledge and expertise required. In-depth knowledge of HVAC systems and advanced design principles enables professionals to optimize energy efficiency and performance in buildings.

Imagine you're tasked with replacing an outdated HVAC system in a large office building. An adept professional doesn't just swap out old units for new ones—they conduct a comprehensive analysis of the building's thermal envelope, occupancy patterns, and existing infrastructure. It's about leveraging technologies like variable refrigerant flow (VRF) systems or integrating smart thermostats that adapt to user behavior in real-time.

According to a recent study on energy-efficient building technologies (CWRU Online Engineering, 2024), using such advanced systems can significantly cut down energy consumption while maintaining or even enhancing indoor comfort levels. Therefore, the role demands continuous learning and adaptation to emerging technologies and methods.

Additionally, specialization within the HVAC field opens up avenues for collaboration with architects, engineers, and contractors on intricate projects. This multidisciplinary approach is key to holistic building design. During the retrofitting of historical buildings, HVAC engineers must work closely with conservationists to ensure the modern systems don't compromise the building's integrity.

Collaborations like this foster a more integrated approach to building design, leading to superior outcomes. Every stakeholder brings their expertise to the table, ensuring that the final product is not only functional but also aesthetically pleasing and compliant with various regulatory standards.

Furthermore, engaging in projects that aim to reduce carbon footprints and enhance living conditions adds a layer of purpose to one's professional life.

For aspiring HVAC professionals, delving into specialized areas such as the integration of energy-efficient technologies is essential. Stay informed about advancements in high-performance HVAC systems, which focus on achieving greater energy savings and improved indoor air quality.

According to the Whole Building Design Guide (WBDG, n.d.), implementing high-performance HVAC equipment alongside a whole-building design approach can result in substantial energy savings, sometimes reducing annual energy costs by as much as 40%.

This speaks volumes about the impact that well-informed and expertly implemented HVAC solutions can have.

Understanding the Career Growth Prospects Within the HVAC Industry

Exploring the diverse career paths in the HVAC industry offers a rich tapestry of opportunities that can truly transform your professional life. For someone seeking a solid understanding of the potential in this field, it's important to recognize that an HVAC career can lead to roles ranging from service technicians to business owners.

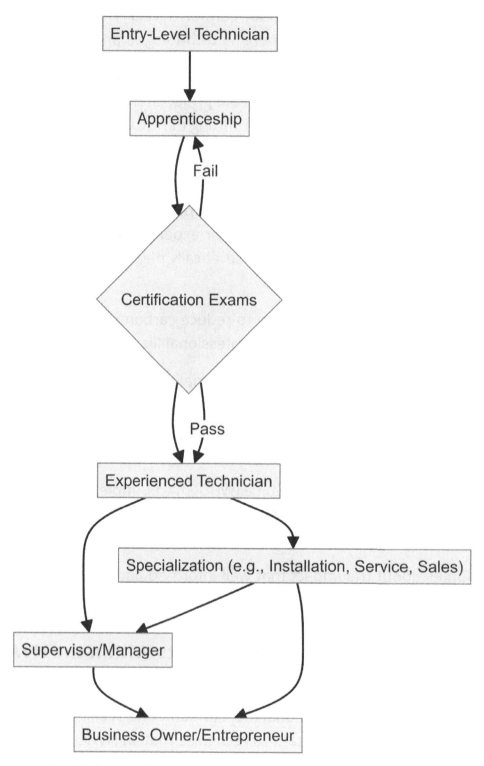

HVAC Career Paths: From Apprentice to Entrepreneur

The HVAC industry is constantly evolving, and staying ahead requires more than just technical expertise. If you're someone who prefers a more hands-on approach, you could start off as a service technician or mechanic. In these roles, you'll get to work on a variety of units, ranging from residential systems to larger commercial ones.

The day-to-day responsibilities of these roles are varied—one moment you're troubleshooting a faulty A.C. unit, the next you're managing ventilation systems in multi-story buildings (Northwestern Tech, 2023). It's precisely this variety that keeps the career interesting and challenging.

But what makes HVAC truly appealing is its robust growth trajectory. According to the U.S. Bureau of Labor Statistics, employment in the HVAC industry is expected to rise by 13% by 2028, significantly outpacing the average for all occupations (admin, 2022). This growth is driven largely by technological advancements and an increasing emphasis on energy efficiency.

Entering the field of HVAC can offer job security and competitive salaries due to the high demand for skilled professionals. In addition, technicians with experience often advance to supervisory or management roles, where they are responsible for overseeing projects and ensuring that quality standards are maintained (RSI, 2021).

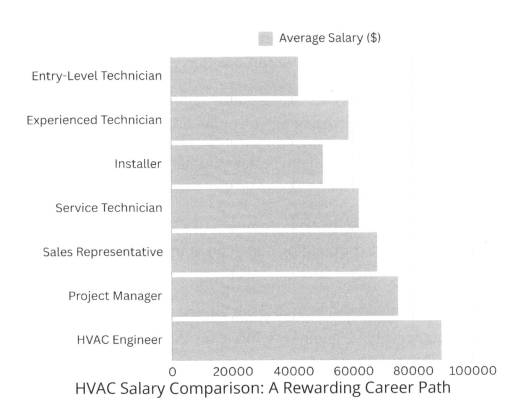
HVAC Salary Comparison: A Rewarding Career Path

If you're someone who values ongoing training and professional growth, the potential rewards can be quite significant. Many people begin their journey by attending trade schools or enrolling in certification programs. They learn important skills such as working with refrigeration and air conditioning systems, heat pumps, and hydronic heating systems. Armed with this knowledge, entry-level technicians can quickly move up the ranks.

Here's how continuous training can help you achieve higher-paying positions and increased responsibilities within HVAC companies:

- Stay updated with emerging technologies. Attending workshops, webinars, or even enrolling in advanced courses can keep you abreast of the latest advancements.

- Seek certifications. Specialized certifications not only enhance your skill set but also make you more attractive to employers.

- Network within the industry. Building relationships with peers, mentors, and industry leaders can open doors to new opportunities.

- Demonstrate leadership. Show initiative by taking on challenging tasks or leading small teams; it showcases your readiness for bigger responsibilities.

Upskilling is another crucial aspect for long-term success in the evolving HVAC sector. As the market shifts toward greener technologies and smarter systems, whoever adapts will find themselves in high demand. Here's how you can position yourself for such advancements:

- Engage with green technology. Learning about solar photovoltaics, wind turbines, and other renewable energy sources can give you an edge in the eco-conscious market.

- Dive into smart HVAC systems. Familiarize yourself with IoT-based HVAC solutions, which promise greater control and efficiency.

- Embrace interdisciplinary skills. Complement your HVAC expertise with knowledge in plumbing, electrical work, or building inspections to become a more versatile technician.

In addition to these technical pathways, there are exciting and unique roles just waiting to be discovered. For instance, HVAC engineers focus on designing mechanical systems and establishing service protocols. They often work closely with clients to plan and coordinate installations, ensuring optimum performance and regulatory compliance (RSI, 2021).

Another exciting role is that of an energy auditor, who assesses HVAC systems for efficiency and recommends improvements—an increasingly vital function as energy costs and environmental concerns grow.

Sales positions also offer lucrative career paths. Sales representatives and engineers require a deep understanding of HVAC systems to recommend appropriate solutions and negotiate contracts. These roles emphasize interpersonal skills and technical knowledge, blending customer service with engineering acumen.

For anyone interested in these diverse pathways, some structured steps can help navigate the career landscape effectively:

- **First,** gain foundational knowledge through accredited training programs.
- **Second,** obtain the necessary certifications and licenses to practice legally.
- **Third,** continually update your skills through professional development opportunities.
- **Fourth,** network aggressively within the industry for mentorship and job opportunities.

To excel in the HVAC field, it's crucial to stay committed to continuous learning and keeping up with industry advancements. If you have dreams of advancing in your career, starting your own business, leading major projects, or becoming an expert in the latest technologies, the secret is to stay competitive and take initiative.

Exploring Entrepreneurial Opportunities in HVAC Business Ventures

Entrepreneurial ventures in the HVAC industry can include starting a contracting business, HVAC consulting firm, or energy efficiency solutions provider. Each of these paths not only provides varied opportunities but also caters to different market needs and personal interests.

Starting an HVAC contracting business involves providing installation, maintenance, and repair services to residential and commercial clients. This path offers the chance to build lasting relationships with customers while ensuring their heating and cooling systems run smoothly.

The key to success here is maintaining a high level of technical competency and customer satisfaction. To get started, you'll need to obtain necessary certifications and licenses, which vary by region. Investing time in acquiring these credentials will ensure you are well-prepared to meet legal and safety standards.

An HVAC consulting firm focuses on providing expert advice to other businesses and homeowners on optimal system designs, upgrades, and efficiency improvements. This venture benefits from strong analytical skills and a vast knowledge of the latest industry trends and technologies.

Professionals in this field often conduct energy audits and recommend systems that align with environmental policies and cost-saving measures. Engaging in continuous education and staying updated on new technologies can make your advice invaluable to clients looking to improve their energy efficiency.

For those interested in the sustainability aspect, becoming an energy efficiency solutions provider could be immensely rewarding. This venture focuses on implementing green technologies like solar heating, geothermal systems, and smart thermostats.

Your role would involve both consulting and hands-on work to enable clients to reduce their carbon footprint. Securing partnerships with manufacturers of eco-friendly products can provide you with the resources you need to offer comprehensive services.

Owning an HVAC business allows individuals to be their own boss, create a unique brand identity, and tailor services to meet market demands. Building a distinct brand requires a deep understanding of your target market and the ability to communicate your value proposition effectively.

Crafting a memorable logo, establishing a professional online presence, and delivering exceptional service are crucial steps in standing out from the competition. By focusing on quality workmanship and excellent customer service, you can create a loyal customer base that drives long-term profitability.

Entrepreneurship requires business acumen, marketing savvy, and customer service skills to build a successful HVAC enterprise. Here's what you can do to achieve this goal:

- Develop a solid business plan outlining your objectives, target market, financial projections, and strategy for growth. This serves as a roadmap to guide your actions and decisions.

- Establish a strong online presence with a professional website and active social media accounts. Engage with potential customers through informative blog posts, DIY tips, and industry news to position yourself as a thought leader.

- Invest in training for both you and your team to stay current with HVAC technologies and best practices. Certifications from recognized institutions can boost your credibility and attract more customers.

- Focus on exceptional customer service by responding promptly to inquiries, offering flexible scheduling, and keeping your clients informed throughout the service process. Satisfied customers are likely to refer your services to others.

- Network with suppliers and other industry professionals to access better pricing on equipment and materials and gain insights into upcoming trends and innovations.

Pursuing entrepreneurship in HVAC requires a blend of technical expertise and business acumen to thrive in a competitive market and achieve financial independence. Balancing the dual roles of technician and business owner means you'll need to be adept at managing time and resources efficiently.

Taking courses in business management or hiring a consultant can provide you with the skills needed to handle administrative tasks, from bookkeeping to employee management.

Additionally, the globalization of the startup ecosystem has been advancing, creating more opportunities for HVAC entrepreneurs. This includes the expansion of overseas markets and the influx of international investments (Boyoung et al., 2018).

Embracing this trend can lead to exploring new avenues for growth and innovation. Collaborating with international partners or adopting proven practices from other countries can help your business stand out.

Being an entrepreneur in the HVAC industry also offers room for creativity and innovation. Whether it's finding new methods to improve energy efficiency or developing a niche market segment, the opportunities are vast and varied. To stay ahead, keep an open mind and be willing to experiment with new ideas and approaches.

In essence, pursuing a career in HVAC is about more than just fixing air conditioning units—it's about committing to excellence, lifelong learning, and making a meaningful impact on society.

As you consider this field, think about how your efforts can improve not only your own life but also the lives of others through enhanced comfort and energy efficiency. With the right training and dedication, you can turn these opportunities into a rewarding and fulfilling career.

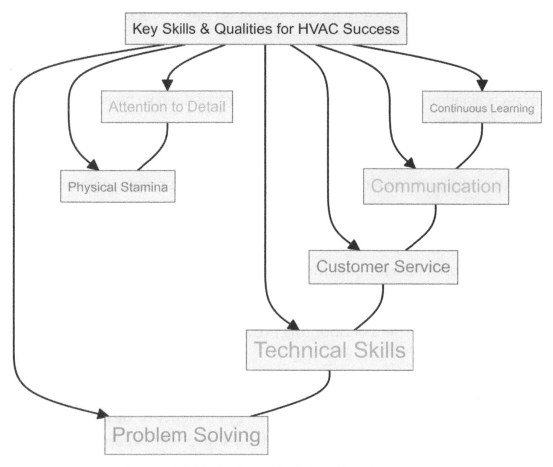

Essential Skills for a Thriving HVAC Career

References

WBDG. (n.d.). High-performance HVAC. Retrieved from https://www.wbdg.org/resources/high-performance-hvac

Austin Career Institute. (2022). What Sort of Jobs and Career Paths do HVAC Technicians Have? [Web log post]. https://austincareerinstitute.edu/blog/hvac/different-types-of-hvac-career-paths/

UEI College. (2023). Is HVAC a Good Career?. UEI College. Retrieved from https://www.uei.edu/trade-school-programs/heating-ventilation-air-conditioning/is-hvac-good-career/

CWRU Online Engineering. (2024). Energy-Efficient Building Technologies: Challenges and Opportunities. Online Engineering Blog. Retrieved from https://online-engineering.case.edu/blog/energy-efficient-building-technologies

Ferriere, T. (2020). 7 Reasons Why Becoming an HVAC Technician Is a Promising Career.... InterCoast Colleges. Retrieved from https://intercoast.edu/articles/hvac-technician-career/

Northwestern Technical College. (2023). HVAC Career Paths - Northwestern Tech. Retrieved from https://northwesterntech.edu/hvac-career-paths/

Ur Rehman, Z., Arif, M., Gul, H., & Raza, J. (2022). Linking the trust of industrial entrepreneurs on elements of ecosystem with entrepreneurial success: Determining startup behavior as mediator and entrepreneurial strategy as moderator. Frontiers in Psychology, 13. https://doi.org/10.3389/fpsyg.2022.877561

RSI. (n.d.). Career advancement options in HVAC. Retrieved from https://www.rsi.edu/blog/hvacr/career-advancement-options-in-hvac/

Kim, B., Kim, H., & Jeon, Y. (2018). Critical Success Factors of a Design Startup Business. Sustainability, 10(9), 2981. https://doi.org/10.3390/su10092981

K, J. (2018). Pros and Cons of Working as an HVAC Technician. Trade School Programs in Chicago. Retrieved from https://www.coynecollege.edu/pros-and-cons-of-working-as-an-hvac-technician/

17
Regulatory Compliance in HVAC

You're about to install a new HVAC system in your home. You've researched the best equipment, planned the layout, and are ready to roll up your sleeves and get to work.

Have you considered the critical role that regulatory standards and codes play in ensuring both the safety and efficiency of your project?

These regulations aren't just bureaucratic hurdles; they are essential guidelines designed to protect you, your home, and anyone who might occupy it in the future.

Building codes serve as the backbone for any HVAC installation, outlining safety measures that must be followed to prevent hazards such as electrical fires or carbon monoxide poisoning.

Proper ventilation requirements ensure your system does not contribute to poor indoor air quality, which can pose serious health risks. These codes also mandate specific procedures for electrical connections and structural integrity, making sure your installations won't fail unexpectedly.

Moreover, permits verify that your plans comply with all local building and environmental regulations, adding another layer of safety and legality to your project. Submitting detailed plans for review may seem tedious, but skipping this step could result in fines, delays, or even dangerous outcomes.

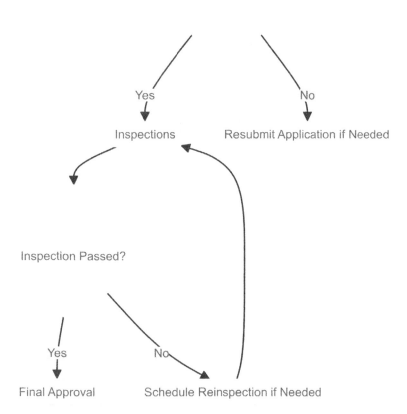

HVAC Permitting Process: A Step-by-Step Guide

Building Codes and Permits for HVAC Installation

Building codes outline safety regulations and quality standards that must be met during HVAC installation to guarantee occupant safety and system efficiency. They're essentially the rulebook ensuring that any HVAC work done meets safety and performance benchmarks.

Think of them as the invisible hand guiding you towards proper procedure and quality outcomes.

For instance, HVAC systems need to meet specific criteria regarding ventilation, electrical connections, and structural integrity. These aren't arbitrary rules; they arise from extensive research and field experience to prevent hazards like fires, carbon monoxide poisoning, or system failures.

Permits are another crucial piece of this regulatory puzzle. These legal documents aren't just bureaucratic red tape; they authorize HVAC installation by verifying that your plans comply with all relevant building regulations and environmental guidelines. Without the right permits, your project could come to a screeching halt, leading to costly delays or even fines.

Permit applications typically involve submitting detailed plans and specifications, which will then be reviewed by local authorities. The goal is to ensure that every aspect, from the layout to the materials used, adheres to established codes.

This oversight helps protect everyone involved—from the installers to the eventual occupants of the building. By securing permits, you're not only following the law but also affirming that your project has been vetted for safety and compliance.

Here's what you can do to avoid costly fines, delays, and risks:

- Make yourself well-acquainted with the specific building codes relevant to your area. Building code requirements can vary significantly between different municipalities.

- Establish a checklist of required permits for your HVAC project. This should include understanding which installations necessitate individual permits.

- Maintain clear and comprehensive records of all permit applications and approvals. Digital filing systems can be particularly useful for tracking these documents.

- Reach out to your local building officials on a regular basis to get clear answers on any confusing requirements or recent changes to the codes..

These steps may seem tedious at first, but they are essential safeguards against unforeseen expenses and project interruptions.

Regular updates on building codes and permits are another indispensable aspect. Regulations are not static; they evolve to incorporate new technologies, materials, and methodologies. Staying informed about these changes ensures that your HVAC projects continue to comply with the latest standards, thereby avoiding any compliance issues down the road.

Keeping up with updates might sound daunting, but there are several straightforward ways to stay current. Subscribing to newsletters from local building departments or professional organizations can provide timely updates.

Moreover, participating in workshops or training sessions offered by these entities can deepen your understanding of new regulations and how they apply to your specific context.

The landscape of HVAC installation isn't just defined by technical know-how but equally by an awareness of the regulatory environment. Being well-versed in building codes and permits forms the backbone of this knowledge, ensuring that every project upholds both safety and efficiency.

To adopt a more evidence-driven policy approach, it's worth exploring how government bodies and private enterprises can collaborate to streamline the process.For instance, online permitting systems backed by robust cybersecurity measures could expedite the approval process while maintaining data integrity and transparency.

Efforts to standardize building codes in larger jurisdictions can help streamline compliance, ensuring that everyone involved can easily follow the rules without any confusion or discrepancies.

Codes ensure our safety and the systems' operational efficiency, while permits confirm that our endeavors align with legal and environmental expectations. Familiarizing yourself with both aspects not only prevents headaches associated with non-compliance but also fosters a safer and more reliable living environment.

EPA Regulations on Refrigerant Handling

One critical aspect of these regulations revolves around the handling of refrigerants, which are substances used in air conditioning systems to cool and dehumidify indoor air. The EPA has established stringent regulations to manage refrigerant handling, primarily driven by environmental responsibility and public health concerns.

EPA regulations aim to curb harmful emissions and promote responsible refrigerant management practices to protect both the environment and public health. At the heart of these regulations is the prevention of ozone-depleting substances from reaching the atmosphere.

Chlorofluorocarbons (CFCs) and hydrochlorofluorocarbons (HCFCs), once widely used as refrigerants, have been significantly phased out due to their harmful effects on the ozone layer. When released into the atmosphere, these substances rise to the stratosphere and break down under ultraviolet light, releasing chlorine or bromine atoms that deplete the ozone layer, which protects life on Earth from harmful UV radiation (EPA et al., 2015).

Proper handling, storage, and disposal of refrigerants are critical to prevent environmental pollution and comply with EPA guidelines. Improper management can lead to unintentional release into the atmosphere, posing significant risks.

Overlap Between EPA Regulations and OSHA Guidelines in HVAC Work

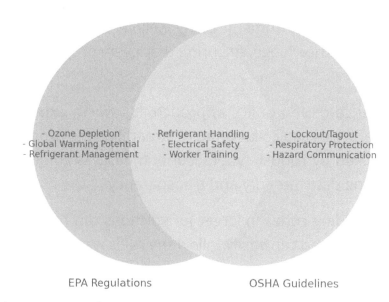

EPA and OSHA: Overlapping Regulations for HVAC Safety and Environmental Protection

To adhere to EPA's comprehensive guidelines:

- Handle refrigerants with care, ensuring no venting occurs during maintenance.

- Store refrigerants in certified, leak-proof containers.

- Dispose of old refrigerants by recovering them using certified equipment, followed by recycling or safe disposal methods.

These steps align with safeguarding our planet while also complying with regulations intended to keep our air clean and our ecosystems healthy.

HVAC professionals must undergo EPA certification to handle refrigerants legally and ensure compliance with these regulations while protecting their health and the environment. This certification, known as Section 608 Technician Certification, ensures that technicians have the knowledge and skills to handle refrigerants safely.

Here is what you can do to achieve certification:

- Study EPA-approved materials covering the rules and proper techniques for handling refrigerants.

- Pass an examination administered by an EPA-approved certifying organization.

- Continue education to stay updated on any changes or updates to the regulations.

This certification not only safeguards technicians but also instills confidence in homeowners and businesses relying on professional services. Technicians with this credential show dedication to environmental conservation and adherence to safety protocols.

EPA regulations foster a culture of sustainability and eco-consciousness within the HVAC industry, driving innovation towards greener practices.

For instance, there's been a significant shift towards using hydrofluorocarbons (HFCs) and other non-ozone-depleting refrigerants, though these too are being scrutinized for their global warming potential.

Complying with these regulations goes beyond merely following laws—it represents a commitment to adopt sustainable practices that benefit current and future generations. It encourages innovation, prompting manufacturers to develop more efficient and environmentally friendly refrigerant alternatives, such as natural refrigerants like carbon dioxide (CO_2) and ammonia (NH_3), which have minimal environmental impact compared to traditional refrigerants.

One compelling illustration of these principles at work is the Montreal Protocol's influence on refrigerant use worldwide. This international treaty, signed by nearly all countries, aims to phase out substances that deplete the ozone layer.

History of EPA Refrigerant Handling Regulations

1990s			2000s	2010s		2020s	
The Clean Air Act Amendments of 1990	EPA's Section 609 Certification Program for technicians	Montreal Protocol	EPA's Significant New Alternatives Policy (SNAP) Program	EPA's Final Rule	Amendment to the Montreal Protocol	EPA's AIM Act	Adoption and Implementation of Natural Refrigerants
Phaseout of CFCs and HCFCs	1992	Accelerated Phaseout of HCFCs	Introduction of HFCs	Decrease in production and import of HCFC-22 and HCFC-142b	Kigali Amendment to phase down HFCs	Regulation for HFC Reduction	2021
1990		1997	2010	2015	2016	2020	

Evolution of EPA Regulations on Refrigerant Handling

Because of the concerted efforts under this protocol, the ozone layer is on a path towards recovery, showcasing how global cooperation and stringent regulations can yield positive environmental outcomes.

Adhering to EPA regulations on refrigerant handling is essential for environmental sustainability, regulatory compliance, and professional credibility in the HVAC sector. Homeowners, too, play a vital role in this equation. By choosing certified technicians and prioritizing regular maintenance of their HVAC systems, they contribute to larger efforts in combating climate change and protecting public health.

Furthermore, maintaining well-functioning equipment leads to enhanced energy efficiency, resulting in lower utility bills and extended system lifespans.

If you're thinking about taking on HVAC projects yourself, it's important to know that there are certain tasks, such as recharging refrigerants or making repairs that could release refrigerants, which you need EPA certification for. This ensures that these tasks are performed correctly, safely, and without adverse environmental impacts.

OSHA Guidelines for HVAC Operations Safety

HVAC technicians often work in potentially hazardous conditions, dealing with refrigerants, electricity, and confined spaces. Occupational Safety and Health Administration (OSHA) regulations are not just arbitrary rules; they are essential safeguards crafted from years of empirical data and worker experiences.

First and foremost, OSHA's safety protocols aim to create a secure working environment. They address the unique risks associated with HVAC work by setting standards that help technicians avoid accidents and injuries.

For instance, handling refrigerants can expose workers to toxic substances, which is why OSHA mandates proper training and the use of protective equipment. Electrical work, another common task for HVAC technicians, poses significant electrocution risks. By requiring adherence to electrical safety standards, OSHA ensures that technicians are well-protected against such dangers.

Proper training on OSHA guidelines is crucial for HVAC technicians. This training equips them with the knowledge and skills needed to identify and address safety risks effectively. Here's what you can do to ensure proper training on OSHA guidelines:

- Ensure all new hires undergo comprehensive OSHA training tailored to HVAC-specific hazards.

- Regularly update training materials to reflect the latest OSHA regulations and industry best practices.

- Conduct hands-on training sessions where technicians can practice safety procedures in controlled environments.

- Encourage ongoing education through workshops, seminars, and online courses focused on OSHA compliance.

- Foster an open dialogue about safety, allowing employees to discuss potential hazards and suggest improvements.

Adhering to these guidelines significantly enhances organizational safety practices. When every team member understands and follows OSHA standards, the workplace becomes inherently safer.

This commitment to safety minimizes the likelihood of injuries and incidents, creating a culture of accountability and compliance within HVAC companies. Employees feel valued and protected, knowing their well-being is a top priority.

To foster this culture, organizations need to incorporate OSHA standards into their day-to-day operations. In a business setting, we should make sure to include safety checks in our everyday tasks and remind everyone about the importance of compliance during team meetings. Management should set a good example by consistently showing their dedication to maintaining safety standards. Over time, this fosters an atmosphere where safety becomes instinctive.

Regular OSHA compliance audits and safety training programs are critical for maintaining industry best practices and safeguarding workers' health and well-being.

These audits serve as checkpoints, ensuring that all safety measures are actively in place and effective. Routine training keeps everyone updated on new regulations and reinforces the importance of safety practices.

Here's how you can uphold regular OSHA compliance audits and training programs:

- Schedule periodic OSHA compliance audits to evaluate current safety practices and identify any shortcomings.

- Appoint a dedicated safety officer responsible for conducting these audits and reporting findings to management.

- Develop a corrective action plan to address any issues identified during audits promptly.

- Implement a continuous improvement process, regularly reviewing and updating safety policies based on audit results and employee feedback.

- Organize mandatory refresher training sessions to keep all staff informed about the latest OSHA guidelines and safety techniques.

Creating a secure work environment goes beyond merely following regulations. It's about cultivating a mindset where safety is prioritized at every level. This proactive approach not only prevents accidents but also promotes overall well-being among workers. For homeowners and DIY enthusiasts, understanding these regulations can also provide insight into selecting reputable HVAC service providers who prioritize safety and compliance.

In the context of HVAC systems, this means that cost-cutting measures should never compromise safety standards. Investing in proper training and equipment might seem like a hefty expense upfront, but it pays dividends by preventing costly accidents and ensuring long-term operational efficiency.

Fostering a culture of safety creates a ripple effect. It encourages personal responsibility among workers, empowering them to take charge of their own safety and that of their colleagues. When individuals feel a sense of ownership over their safety practices, compliance becomes more robust and ingrained in the organizational fabric.

For novice technicians entering the HVAC field, understanding OSHA guidelines early in their careers sets a solid foundation for professional growth. It instills a discipline that will guide them through various job roles and responsibilities. Technicians trained in environments where OSHA guidelines are strictly followed tend to carry those practices with them, contributing to safer workplaces across the industry.

Energy Efficiency Regulations in HVAC Design

The impact of energy efficiency regulations on HVAC system designs cannot be overstated, as they set the stage for sustainable technologies and overall energy conservation.

Energy efficiency regulations establish performance benchmarks for HVAC equipment. These benchmarks encourage the use of sustainable technologies aimed at minimizing energy consumption.

By setting these benchmarks, regulators aim to guide people toward making choices that benefit both their wallets and the environment. When HVAC systems meet or exceed these benchmarks, they tend to consume less energy while maintaining the same level of performance, if not improving it.

One significant advantage of complying with energy efficiency standards is the reduction in operational costs for building owners. More efficient HVAC systems spend less time running at full capacity, thereby consuming less energy and saving on utility bills.

Additionally, energy-efficient systems contribute to environmental conservation by lowering carbon emissions, a critical step in combating climate change. When you make the switch to energy-efficient HVAC systems, you're not just saving money; you're also playing a part in preserving our planet.

Integrating energy-efficient practices into HVAC designs offers multiple benefits. Not only does it promote long-term savings, but it also enhances system efficiency and aligns with global initiatives for sustainable development.

To achieve this, here are some steps you can follow:

- Start by evaluating the current efficiency of your HVAC system. Look for areas where improvements can be made without compromising comfort.

- Next, invest in high-efficiency components that meet or exceed regulatory benchmarks. This might involve updating to ENERGY STAR®-certified equipment, which can save you between 10% to 30% annually on your energy bills (U.S. Department of Energy and ENERGY STAR®).

- Consider integrating smart technology, like programmable thermostats and automated ventilation controls, to optimize your system's performance.

- Finally, conduct regular maintenance checks and tune-ups to ensure that your system remains efficient over its lifespan.

Being aware of energy efficiency regulations not only aids in compliance but also opens doors to various incentives, rebates, and tax credits associated with eco-friendly solutions. This is incredibly beneficial for HVAC professionals looking to adopt green technologies.

For instance, many states and energy companies offer financial incentives to encourage the installation of energy-efficient HVAC systems. Here's how you can leverage these opportunities:

- Keep updated on local and federal energy efficiency programs. Websites like ENERGY STAR® provide comprehensive information about available incentives and how to apply for them.

- Consult with HVAC professionals who specialize in eco-friendly solutions to identify all possible rebates and tax credits you qualify for.

- Ensure that all installations and upgrades are documented properly to make claiming incentives straightforward.

- Use the savings from these incentives to further invest in other energy-saving measures, like better insulation or renewable energy sources.

The practical upshot of adhering to energy efficiency regulations is that it leads to cost savings, environmental stewardship, and enhanced system performance, aligning perfectly with sustainable practices for a greener future. High-performance HVAC systems are not just about meeting today's needs but also preparing for a more sustainable tomorrow. Systems that prioritize human welfare while still supporting economic growth showcase this balance beautifully.

It's also worth noting that heating and cooling account for about half of a typical home's energy usage (U.S. Department of Energy and ENERGY STAR®). Hence, by focusing on high-performance HVAC systems, we tackle one of the largest segments of household energy consumption.

This makes adhering to energy efficiency regulations not just a matter of policy compliance but a smart economic choice for individuals and businesses alike. By adhering to these regulations, investing in advanced technology, and using available incentives, we can create HVAC systems that are not only cost-effective but also significantly kinder to our planet.

And who knows? The next great advancement in HVAC technology might come from someone just like you, dedicated to balancing personal responsibility with the need for a robust safety net for those who fall on hard times.

However, some readers might worry about the seeming complexity of these regulations and the time it takes to secure necessary permits. These concerns are valid but manageable.

Familiarizing yourself with the specific requirements relevant to your area and maintaining clear records can alleviate much of this stress. Regular consultations with local building officials can provide clarity and keep you updated on any changes.

Let's end on a thoughtful note: staying informed and compliant with building codes and permits is not just about ticking off boxes. It's about engaging in practices that ensure safety, efficiency, and peace of mind.

References

U.S. Department of Energy. (n.d.). Tech Solutions. https://rpsc.energy.gov/tech-solutions/hvac

Asim, N., Badiei, M., Mohammad, M., Razali, H., Rajabi, A., Haw, L. C., & Ghazali, M. J. (2022). Sustainability of Heating, Ventilation and Air-Conditioning (HVAC) Systems in Buildings—An Overview. International Journal of Environmental Research and Public Health, 19(2), 1-16. https://doi.org/10.3390/ijerph19021016

Occupational Safety and Health Administration. (n.d.). Ventilation. OSHA. Retrieved from https://www.osha.gov/ventilation

Illinois Capital Development Board. (2024). Building codes and regulations. Retrieved from https://cdb.illinois.gov/business/codes/buildingcodesregulations.html

US EPA, OAR. (2015). Managing refrigerant in stationary refrigeration and air-conditioning equipment. Retrieved from https://www.epa.gov/section608/managing-refrigerant-stationary-refrigeration-and-air-conditioning-equipment

Occupational Safety and Health Administration. (n.d.). Ventilation standards. Retrieved from https://www.osha.gov/ventilation/standards

Environmental Protection Agency. (2016). Regulatory Updates: Section 608 Refrigerant Management Regulations. EPA. https://www.epa.gov/section608/regulatory-updates-section-608-refrigerant-management-regulations

RSI Education. (n.d.). How Does OSHA Affect the HVAC Industry? Retrieved from https://www.rsi.edu/blog/hvacr/osha-affect-hvac-industry/

Environmental Protection Agency. (2015). Homeowners and consumers: Frequently asked questions. [Webpage]. https://www.epa.gov/ods-phaseout/homeowners-and-consumers-frequently-asked-questions

City of Chicago. (2023). Chicago Building Code Online. Chicago Construction Codes. Retrieved from https://www.chicago.gov/content/city/en/depts/bldgs/provdrs/bldg_code/svcs/chicago_buildingcodeonline.html

Cook County. (n.d.). Building permits. Cook County. Retrieved from https://www.cookcountyil.gov/service/building-permits

18

Case Studies in HVAC

Let's dive into some real-life examples that demonstrate how residential HVAC retrofits can be a game-changer. Discover the incredible impact that modern technologies, such as smart thermostats and high-efficiency units, can have on enhancing the comfort and energy efficiency of your home. Let's dive deeper into the significance of using proper insulation and sealing techniques to improve system performance.

You'll get a practical understanding of how to apply HVAC concepts in everyday situations with these detailed examples and actionable insights. This will help you make informed decisions about your home's heating and cooling needs.

The Process and Benefits of a Residential HVAC Retrofit Project

Analyzing an existing HVAC system is akin to a detective working on a case, where every clue might lead to enhanced energy efficiency and superior comfort for your home.

When embarking on a residential HVAC retrofit project, the first step involves thoroughly examining the current system. Imagine a house with cold drafts in winter or hot spots in summer—these are clear indicators that something's amiss.

By identifying malfunctioning components, leaky ducts, or outdated units, homeowners can pave the way to a more comfortable and cost-effective living environment.

Implementing modern technologies holds tremendous potential. Smart thermostats, for instance, are game changers. They don't just allow for remote control of heating and cooling but can learn patterns and adjust settings automatically, optimizing energy use without sacrificing comfort.

Upgrading to energy-efficient HVAC units can also yield long-term savings. According to the U.S. Department of Energy, choosing high-efficiency equipment can cut electricity use by up to 50% for electric systems and 10% for gas systems (U.S. Department of Energy, 2023). This isn't merely about economics; it's about balancing responsible energy consumption with personal convenience.

Here's a step-by-step guide to help you achieve your goal:

- Begin by researching different models of smart thermostats and energy-efficient HVAC units.

- Compare the features and functionalities to find those best suited for your home's particular needs.

- Consult with professionals to ensure proper installation and integration of these modern technologies into your existing setup.

- Take advantage of potential rebates or incentives offered by utilities or governmental programs to offset the initial investment costs.

Proper insulation and sealing techniques play an indispensable role in any HVAC retrofit. If you think of your home as a vessel, insulation and sealing act like its outer hull, preventing precious conditioned air from leaking out. A substantial percentage of energy loss occurs through poorly sealed ducts and uninsulated spaces.

According to ENERGY STAR®, approximately 20% of conditioned air is lost while moving through a typical home's duct system due to air leakages (ENERGY STAR®, 2023). Improved insulation not only helps in maintaining the desired temperature but also ensures the HVAC system doesn't have to work overtime, thereby prolonging its life.

Here's a step-by-step guide to help you achieve your goal:

- Perform a comprehensive inspection of your home's insulation status and ductwork.

- Use tools like thermal imaging cameras to pinpoint areas with poor insulation or significant leaks.

- Seal leaks with appropriate materials such as mastic sealant or metal-backed tape for ducts.

- Consider adding insulation to attics, walls, and crawl spaces to further enhance energy retention.

Lastly, regular maintenance and monitoring are paramount to making sure your retrofitted system remains effective over time. Neglecting routine upkeep can negate all the benefits accrued from the retrofit. Filters need changing, ducts need cleaning, and systems require periodic check-ups.

Just as you wouldn't skip regular medical check-ups, the same principle applies here. Routine maintenance ensures small issues don't balloon into larger, costly problems and keeps the system operating at peak performance.

The Impact of Professional HVAC Maintenance (Before & After)

Here's a step-by-step guide to help you achieve your goal:

- Schedule semi-annual professional inspections to evaluate and maintain the system.

- Regularly change air filters, typically every 1-3 months, depending on usage and manufacturer recommendations.

- Keep an eye on energy bills; unexpected increases could signal inefficiencies needing attention.

- Use DIY practices like checking the thermostat's accuracy and ensuring vents are clear of obstructions.

The journey starts with understanding your current HVAC system's strengths and weaknesses. By incorporating smart technologies and energy-efficient units, you can achieve long-term cost savings and enhance your overall comfort. Insulation and sealing play a crucial role in protecting the effectiveness of a top-notch HVAC system.

Lastly, consistent maintenance habits fortify these improvements, leading to sustainable living environments.

Exploring a Commercial HVAC System Upgrade Case Study to Showcase Scalable Solutions

When confronted with upgrading a commercial HVAC system, we enter a realm where multiple variables come into play. These upgrades aren't just about replacing old components with new ones; they're comprehensive projects involving considerations such as zoning, occupancy patterns, and specialized ventilation requirements tailored to the nature of each business.

Imagine a large office building that is humming with activity during work hours but relatively silent after-hours and on weekends. It's essential to design an HVAC system that can adapt to these shifting needs without wasting energy or compromising comfort.

Turning our attention to retrofitting existing systems, this process yields significant benefits. Energy-efficient components save money on utility bills and reduce environmental impacts. For businesses, every dollar saved on operational costs can make a substantial difference.

Additionally, using greener technology aligns with social responsibility goals—a win-win scenario. Here's how:

- Assess your existing system to identify outdated or inefficient components
- Choose replacements that are proven to boost efficiency and lower energy consumption
- Work with professional technicians who understand the nuances of your specific needs
- Consider grants or incentives available for energy-efficient upgrades

Next, let's delve into project management. The success of an HVAC upgrade hinges largely on effective coordination and communication among stakeholders - from engineers and contractors to facility managers and staff. Planning meticulously reduces disruptions in the business environment and ensures everyone is aligned towards the same goals. Key steps include:

- Setting clear objectives and timelines
- Designating roles and responsibilities for all team members involved

- Continuously monitoring progress and adapting plans to address unforeseen challenges

- Maintaining transparent and open lines of communication to keep everyone informed

Once the upgrade is complete, it doesn't mean the job is done. Monitoring and tracking performance metrics post-upgrade are crucial to ensure the system operates at its optimal efficiency. By carefully observing data related to energy use, temperature consistency, and overall cost savings, valuable insights emerge that can be used to fine-tune the system further.

Here's what you can focus on:

- Regularly track energy consumption and compare it against historical data

- Monitor the indoor climate to verify that the system maintains desired comfort levels

- Use smart sensors and IoT technologies to gather real-time data for more accurate analysis

- Schedule routine maintenance checks to preemptively address any issues

From this multi-faceted approach, one important lesson becomes clear: the utmost importance lies in customization and scalability. Every commercial establishment has its own set of needs and limitations, making each one unique. Therefore, a solution that works for everyone is not sufficient. Instead, it is both practical and essential to customize HVAC systems in order to meet specific business needs and regulatory standards. This ensures long-term satisfaction and efficiency.

Here's an example: a hospital needs different zones for various departments like surgery, radiology, and patient recovery rooms. Each zone will have distinct heating, cooling, and ventilation requirements. Tailoring the system to accommodate these differences while ensuring compliance with health regulations demands both careful planning and skillful execution.

Scalability matters particularly for growing businesses. A well-designed HVAC system should allow for an easy expansion to handle an increasing number of occupants or a larger workspace without necessitating a complete overhaul. Flexibility in design can save both time and money down the line.

Let's consider a case study to illustrate this further.

A mid-sized tech startup is housed in an older building. Initially, their HVAC system was a patchwork of aging units that struggled to keep up, especially when the team doubled in size within a year. Realizing the importance of upgrading, they opted for a zoned HVAC system featuring energy-efficient components like VRF units.

The implementation phase was meticulous. They started by evaluating their existing setup and pinpointing inefficiencies. They coordinated with HVAC professionals to install the new system while making sure to minimize downtime. And after installation, continuous monitoring revealed that their energy bills dropped by 30%, and employee comfort levels skyrocketed, thanks to a more consistent indoor climate.

This example underscores a simple truth: investing in modern, efficient HVAC solutions pays off in both tangible and intangible ways. The tech startup not only reaped financial rewards but also saw productivity gains and higher employee satisfaction—a testament to the value of taking a considered, evidence-based approach to HVAC upgrades.

Think of it as a holistic strategy that balances economic growth with human welfare. In an age of increasing awareness around energy efficiency and sustainability, such balanced approaches are not just preferable—they're imperative. As we continue navigating evolving market forces and governmental policies, making informed choices grounded in empirical evidence will always light the way forward.

Remember: successful HVAC upgrades blend smart technology with sound strategies, always prioritizing the people they serve.

Highlighting the Design Principles of Energy-Efficient HVAC Systems for Sustainable Buildings

Integrating renewable energy sources such as solar panels or geothermal heat pumps can significantly reduce the environmental footprint of HVAC systems in sustainable buildings. Think about how harnessing the power of the sun to generate electricity for your HVAC system can not only cut down on your utility bills, but also contribute to a greener planet.

Let's take a real-world example: The Bullitt Center in Seattle, often regarded as one of the greenest commercial buildings in the world, uses solar panels extensively. These panels provide almost all the electricity needed to run the building's HVAC system. By installing solar panels, homeowners and businesses alike can drastically decrease their reliance on non-renewable energy sources, promoting both economic savings and environmental sustainability.

Proper sizing and configuration of HVAC equipment based on building loads and usage patterns are crucial in achieving energy efficiency goals. Imagine wearing shoes that are too big or too small; they would either be uncomfortable or inefficient, causing you to tire out quickly. Similarly, an improperly sized HVAC system will either consume excessive energy (if oversized) or fail to maintain comfort levels (if undersized).

Here are the steps you can take:

- Conduct a thorough load calculation considering the building's dimensions, insulation, and local climate.

- Ensure that your HVAC professional assesses the unique usage patterns and peak load times of the building.

- Avoid generic equipment sizes; customize the HVAC components to meet specific needs.

- Regularly review and adjust the system settings to accommodate any changes in usage or building structure.

Using advanced control systems and smart technologies can improve energy management and make occupants more comfortable in sustainable HVAC designs. It's like having a personal assistant who understands your temperature preferences and adjusts the thermostat accordingly while you're at work or asleep.

Modern HVAC systems equipped with smart thermostats and sensors can do just that. They can modulate heating and cooling based on occupancy, outside weather conditions, and personal settings.

Here is what you can do:

- Integrate smart thermostats that allow remote control via smartphone apps.

- Use sensors that detect occupancy levels and adjust the HVAC settings automatically.

- Leverage machine learning algorithms for predictive maintenance, reducing downtime and energy wastage.

- Employ real-time monitoring systems to track energy use and optimize performance continually.

Regularly evaluate and optimize HVAC systems in green buildings to ensure their long-term sustainability. Think of your HVAC system like a marathon runner. Even the best runners need regular check-ups and fine-tuning to perform at their best.

Regular inspections are essential for identifying and addressing small issues before they escalate into major problems, which ultimately helps to maintain the system's durability and effectiveness.

- Schedule bi-annual inspections with a certified HVAC technician to evaluate the system's condition.

- Monitor key performance indicators like energy consumption and temperature regulation efficacy.

- Implement periodic cleaning routines for filters, ducts, and coils to maintain airflow and efficiency.

- Adopt a proactive approach to repairs and upgrades, replacing outdated components with newer, more efficient models.

To bring it all together, creating energy-efficient and environmentally friendly buildings necessitates a holistic approach that integrates architectural design, HVAC system configuration, and sustainable practices.

The Joyce Centre for Partnership and Innovation in Canada serves as a prime model by combining net-zero energy principles with state-of-the-art HVAC solutions. By incorporating passive solar design, extensive use of daylight, and renewable energy sources, this building stands as a testament to how sustainable practices can be seamlessly integrated into modern architecture.

Similarly, at a residential level, employing these principles can drastically improve a home's energy efficiency and comfort.

- Orienting your home to maximize natural light reduces the need for artificial lighting and heating during the day.

- Using high-performance windows and insulation can maintain comfortable indoor temperatures with less strain on the HVAC system.

- Planting deciduous trees around your property can provide shade during the summer and allow sunlight during the winter, effectively reducing heating and cooling demands.

Advanced control systems, such as smart thermostats, can further refine energy usage by adapting to your daily routine and adjusting settings accordingly. Or, consider how sensors can detect when rooms are unoccupied and lower heating or cooling efforts, thereby conserving energy without sacrificing comfort.

Ultimately, the fusion of good architectural design, precise HVAC system integration, renewable energy resources, and smart technology culminates in a building that's not only energy-efficient but also aligned with broader sustainable practices. Whether it's a commercial skyscraper or a suburban home, adopting these principles ensures that we move towards a future where buildings are not just structures but active participants in our global effort to reduce environmental impact.

Present HVAC Troubleshooting Case Studies and Solutions for Real-World Application

When you're faced with the challenge of diagnosing HVAC issues, having a systematic approach in your toolkit can be invaluable. Let's dive into some real-world case studies to understand how these procedures work and, more importantly, to build a practical skill set for tackling common problems.

First up, identifying common HVAC issues like airflow restrictions or refrigerant leaks can seem daunting. But you can simplify the process through systematic diagnostic procedures.

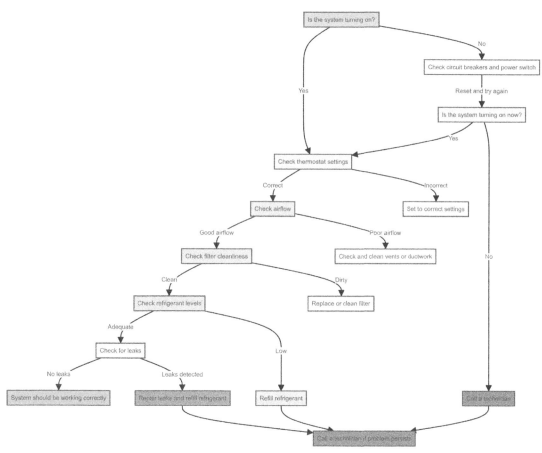

Troubleshooting Common HVAC Issues: A Step-by-Step Guide

Consider a scenario where a homeowner notices reduced airflow from their vents. The initial reaction might be to call a technician immediately, but let's break it down first:

- Start by checking the air filter. A clogged air filter can significantly reduce airflow, affecting system efficiency.

- Inspect the ductwork for visible obstructions or damage. Closed or blocked vents are often easy to spot and fix.

- After addressing these basic checks, use a pressure gauge to measure the system's static pressure. This gives you insights into potential blockages or leaks within the system.

These steps not only help pinpoint the issue but also save time and money. If you're still unsure, bringing in a professional is always a wise move, but understanding these basics can be very empowering.

Now, let's look at resolving some technical problems, such as electrical component failures and system malfunctions. Imagine you're dealing with an HVAC system that doesn't turn on. Here's what you can do to troubleshoot:

- Check the thermostat settings. Ensure it's set to the correct mode and temperature.

- Inspect circuit breakers and fuses. Reset any tripped breakers and replace blown fuses.

- Look at the wiring connections. Loose or corroded wires can interrupt the power supply.

By following these steps, you'll systematically narrow down the source of the problem. Often, it's something straightforward, but if it's more complex, these preliminary checks can help guide you or a technician to the solution faster.

While diving into these technical aspects, it's crucial to emphasize the importance of safety. HVAC systems involve electricity, moving parts, and potentially harmful refrigerants, so observing safety protocols is non-negotiable. Here's a simple guideline to keep in mind:

- Always turn off the power before inspecting or repairing any components.

- Wear protective gear, such as gloves and safety goggles, to protect against electrical shocks or debris.

- Be aware of local regulations and guidelines regarding HVAC repairs and installations to avoid legal issues and ensure your safety.

These measures prevent accidents and ensure that the troubleshooting process is as safe as possible.

Another key aspect of maintaining HVAC efficiency is preventive maintenance. Regular check-ups can help avoid recurring issues. This routine can keep your system running smoothly:

- Replace air filters every one to three months, depending on usage and indoor air quality.

- Clean the evaporator and condenser coils annually to maintain efficient heat exchange.

- Lubricate moving parts, such as fan motors and bearings, to reduce friction and wear.

- Check refrigerant levels and top up as necessary to ensure optimal cooling performance.

Implementing these practices proactively prevents minor issues from escalating into major repairs. Plus, it extends the life of your HVAC system, keeping it more efficient and reliable.

HVAC Maintenance Schedule: A Visual Guide

Let's remember the overarching goal: equipping you with practical knowledge and problem-solving skills to handle a broad range of HVAC challenges. Mastering these will give you the confidence to diagnose and troubleshoot common issues effectively while ensuring the safety and longevity of your system.

Take, for instance, an incident involving a malfunctioning central air conditioner. A homeowner observed that several rooms weren't cooling despite the system running continuously. Upon investigation, here's what was found (Energy.gov, n.d.):

- The unit had been improperly installed, resulting in leaky ducts and reduced airflow.

- The refrigerant charge wasn't aligned with the manufacturer's specifications, affecting both performance and efficiency.

- Unqualified service technicians had previously added refrigerant to an already full system, worsening the existing issues.

To address these points, here's what can be done:

- Properly seal all ductwork to eliminate leaks and improve airflow.

- Use manufacturer guidelines to adjust the refrigerant charge accurately.

- Hire certified technicians who follow best practices for HVAC installation and maintenance to ensure long-term reliability.

Such real-world examples highlight the critical need for systematic approaches and adherence to best practices. By applying these methods, you can achieve a sustainable and equitable approach to HVAC management.

The technical details might seem daunting. Some things need to be evaluated and handled by professionals, while others can be done on your own. It's important to find the right balance between DIY practices and knowing when to bring in experts. By taking a comprehensive approach, we prioritize safety and optimize the effectiveness of your HVAC investments.

References

U.S. Department of Energy. (n.d.). HVAC. Retrieved from https://rpsc.energy.gov/tech-solutions/hvac

Energy.gov. (n.d.). Common Air Conditioner Problems. Retrieved from https://www.energy.gov/energysaver/common-air-conditioner-problems

City College of New York. (2018). SUS 7600B Design of Mechanical Systems for Sustainable Buildings. Retrieved from https://www.ccny.cuny.edu/sustainability/sus-7600b-design-mechanical-systems-sustainable-buildings.

WBDG. (n.d.). Optimize Energy Use. Retrieved from https://www.wbdg.org/design-objectives/sustainable/optimize-energy-use

Sahoh, B., Kliangkhlao, M., & Kittiphattanabawon, N. (2022). Design and development of Internet of Things-driven fault detection of indoor thermal comfort: HVAC system problems case study. Sensors (Basel, Switzerland), 22(5), 1925. https://doi.org/10.3390/s22051925

Montgomery County, Pennsylvania. (n.d.). Renewable energy and energy conservation site and building design. Montgomery County, PA. Retrieved from https://www.montgomerycountypa.gov/DocumentCenter/View/5051/Renewable-Energy_Energy-Conservation-Site-and-Building-Design

Walker, I. S. (2003). Best practices guide for residential HVAC Retrofits. Lawrence Berkeley National Lab. (LBNL), Berkeley, CA (United States). https://doi.org/10.2172/824856

19

The Future of HVAC

With the rapid advancement of smart technology, HVAC systems are now offering more than just convenience. They are also delivering substantial energy savings and improved indoor air quality.

But, keeping up this perfect atmosphere comes with its fair share of difficulties. Conventional HVAC systems can be inefficient, resulting in greater energy usage and higher expenses. These systems might have difficulty adjusting to different conditions, like varying occupancy levels or outdoor weather. As a result, they may not always provide consistent comfort and could end up using more energy than necessary.

In addition, older HVAC systems often need reactive maintenance. They tend to break down without warning, leading to unexpected inconveniences and expenses. This approach is not a good investment and will not be able to be maintained in the long term.

Let's explore how automation and control systems are making HVAC operations smarter, allowing for real-time adjustments based on environmental factors.

Advancements in HVAC Automation and Control Systems

Advancements in automation lead to increased efficiency and reduced energy consumption in HVAC systems. They can modulate heating and cooling output depending on the time of day, occupancy levels, and even weather forecasts. Research supports that these automated adjustments result in significant energy conservation (Advancements in HVAC-R Technology, 2017).

The beauty of it is that this enhanced efficiency doesn't just benefit your wallet but also contributes positively to reducing our carbon footprint.

Moving beyond energy savings, let's explore how smart technology is reimagining the way we interact with HVAC systems.

With remote monitoring and control capabilities, users can now manage their HVAC settings from virtually anywhere with an internet connection. Imagine being able to adjust your home's temperature while you're commuting back from work, ensuring you arrive to a comfortable environment. Here's what you can do to take advantage of this convenience:

- Invest in a Wi-Fi-enabled thermostat compatible with your existing HVAC system.
- Download and set up the associated mobile app on your smartphone or tablet.
- Use the app to monitor current system performance and make necessary adjustments remotely.
- Set up scheduling features within the app to automate temperature changes based on your daily routine.

This level of accessibility transforms HVAC management from a manual, often overlooked task into an effortlessly integrated part of modern living (jenks2026, 2024).

Another remarkable impact of automation in HVAC systems is improved predictive maintenance. Traditional maintenance schedules often rely on reactive approaches—waiting for something to break before fixing it.

However, automated systems equipped with machine learning algorithms can anticipate issues before they cause significant downtime. Here's how you can implement predictive maintenance:

- Begin by upgrading your HVAC system with sensors capable of tracking key performance metrics like airflow, pressure, and temperature.
- Use a centralized software platform to gather data from these sensors continuously.
- Analyze the data to identify patterns indicative of potential faults or inefficiencies.
- Schedule timely maintenance activities based on the insights derived from the analysis.

By addressing problems proactively, not only do you reduce the likelihood of unexpected breakdowns, but you also extend the lifespan of your equipment, ultimately saving both time and money.

Turning our attention to another critical aspect, the adoption of advanced control systems significantly enhances indoor air quality and comfort.

Today's HVAC technologies are smarter than ever—they don't merely maintain a set temperature; they actively manage humidity levels, filter out pollutants, and ensure proper ventilation. This comprehensive approach to climate control means that occupants enjoy a more comfortable and healthier living environment.

Integrated systems can dynamically adjust to minimize allergens and contaminants, making them especially beneficial for individuals with respiratory conditions or allergies.

Reflecting on these advancements, it's clear that the intersection of technology and HVAC systems offers tremendous benefits. Enhanced efficiency, convenience through smart technology, predictive maintenance, and superior indoor air quality collectively illustrate how far we've come. At the heart of these innovations lies the core belief that human welfare must always take precedence over economic growth.

While these technological strides undeniably drive economic benefits, their ultimate value is measured in the improved quality of life they provide.

The evolving landscape of HVAC technology isn't just about future-proofing buildings; it's about creating environments where people thrive. And while the promise of automation and smart systems is compelling, it's equally important to acknowledge the role of skilled technicians who maintain and troubleshoot these complex systems.

The blend of human expertise and cutting-edge technology makes for resilient and adaptive solutions. As this field continues to evolve, staying informed and being open to adopting new technologies will be key to reaping the full benefits of these advancements.

Impact of IoT and AI on the HVAC Industry

Exploring the impact of IoT and AI on the HVAC industry brings us into an exciting realm where technology meets comfort seamlessly. It's not just about making our homes more efficient; it's about integrating data-driven decisions that can lead to significant improvements in how we experience indoor environments. Let's dive into what this futuristic approach entails.

When you think of IoT connectivity, imagine a network of smart devices constantly communicating with each other to make informed decisions. In the context of HVAC systems, IoT takes center stage by enabling real-time monitoring and control. Sensors embedded within these systems collect vast amounts of data—temperature, humidity, energy consumption, and even occupancy patterns.

This treasure trove of information allows for data-driven decision-making, which can revolutionize how we manage our heating, ventilation, and air conditioning units.

For homeowners and DIY enthusiasts looking to leverage IoT connectivity in their HVAC systems, it's essential to start with ensuring your system is compatible with IoT devices. Investing in smart thermostats, sensors, and controllers that can connect to the internet is a good first step.

Next, setting up a central hub or using a cloud-based platform can help streamline data collection and analysis. The beauty of IoT is in predictive maintenance.

Imagine a scenario where your HVAC system alerts you to a potential issue before it becomes a costly repair. By continuously monitoring performance metrics, these smart systems can predict when components are likely to fail, allowing for timely intervention.

Here's what you can do to harness the power of IoT for your HVAC:

- Ensure your HVAC system is compatible with IoT devices.
- Invest in smart thermostats, sensors, and controllers.
- Set up a central hub or use a cloud-based platform for data management.
- Monitor performance metrics regularly to stay ahead of maintenance issues.

Moving forward to the role of AI applications, let's consider how artificial intelligence can optimize HVAC performance based on real-time data and user preferences. AI algorithms analyze the data collected from IoT-enabled devices to find patterns and adjust automatically.

If the system detects a room is consistently warmer than desired during certain hours, it can adjust the cooling schedule to match the user's preference without any manual input. This not only enhances comfort but also ensures energy efficiency by preventing unnecessary heating or cooling.

To integrate AI for optimal HVAC performance, you should look into systems that support machine learning capabilities. These systems can learn from past data, adapt to changing conditions, and even forecast future needs.

It's like having a personal assistant for your HVAC, fine-tuning settings to ensure maximum comfort while minimizing energy usage. Pairing AI with IoT means you can set parameters and preferences, then let the system do the heavy lifting, adjusting as needed based on real-time feedback.

Now, let's explore how the integration of IoT and AI technologies results in smarter, self-learning HVAC systems.

This combination creates an ecosystem where devices don't just follow pre-set instructions but adapt and evolve. Imagine your HVAC system learning your daily routine—when you leave for work, when you return, your preferred temperature settings at different times of the day—and adjusting accordingly. This intelligent operation leads to increased efficiency, as the system operates only when necessary and at optimal levels.

While standing back and letting technology take over might sound daunting, it actually simplifies the user experience. No longer do you need to tinker with the thermostat or worry about whether you've left the AC running unnecessarily. The system learns and adapts, ensuring both comfort and efficiency.

For those inclined towards DIY projects, exploring options in smart HVAC systems can be both a learning experience and a practical investment in home improvement.

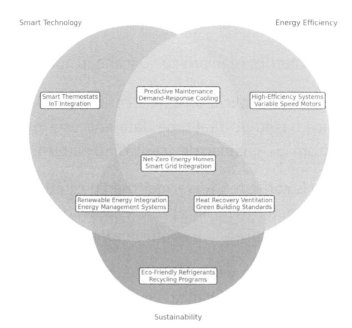

The Convergence of Smart Technology, Energy Efficiency, and Sustainability in the Future of HVAC

Looking ahead, we can see how AI is driving advancements in fault detection and automatic system adjustments. Imagine a future where your HVAC system can anticipate malfunctions and fix them on its own.

These advancements suggest a future where the system takes care of itself, handling tasks like updating software, adjusting settings, and ordering replacement parts as necessary.This would epitomize convenience and reliability, moving beyond mere reactive maintenance to proactive care.

Future implementations could see HVAC systems integrated with home automation platforms, providing seamless interactions across all smart home devices. It's an exciting prospect, blending convenience with sustainability—a crucial element as we prioritize human welfare alongside economic growth.

By leveraging IoT and AI, not only can homeowners maintain ideal indoor climates effortlessly, but you can gain new insights and skills, opening doors to innovative applications in HVAC management.

The future is clearly one where technology doesn't just serve us but anticipates and adapts to our needs, creating environments that are not only smart but also kind to both people and the planet.

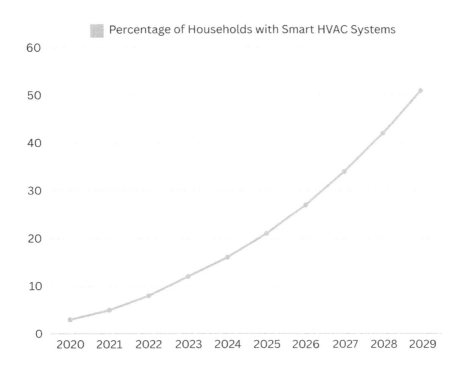

Projected Growth of Smart HVAC Systems in Households

Green HVAC Initiatives and Sustainability Practices

Sustainable HVAC practices focus on reducing energy consumption and environmental impact through efficient systems and renewable energy integration. To achieve this, homeowners and technicians should consider various strategies that blend technological innovation with common-sense adjustments:

- Start by conducting an energy audit of your home or building to understand where most of the energy consumption occurs. This will help pinpoint inefficiencies and areas needing improvement.

- Look into upgrading to high-efficiency HVAC systems if your current setup is outdated. Modern units are designed to use less energy while providing superior performance.

- Incorporate programmable thermostats to optimize heating and cooling schedules based on occupancy. This small change can significantly reduce energy usage without compromising comfort.

- Ensure your HVAC system undergoes regular maintenance. Clean filters, ducts, and vents can vastly improve system efficiency and longevity.

- Explore adding insulation or improving existing insulation. This helps maintain indoor temperatures, reducing the load on your HVAC system.

These steps work together to create a more energy-efficient environment that not only reduces utility bills but also contributes positively to the planet.

Green HVAC initiatives promote environmentally friendly refrigerants and technologies to minimize carbon footprint. Traditional refrigerants often have high global warming potential (GWP), leading to increased environmental impact.

Fortunately, advancements in refrigerant technology provide more sustainable options. For instance, using hydrofluoroolefins (HFOs) can significantly decrease your system's GWP. HFOs are designed to break down more quickly in the atmosphere, meaning they contribute less to long-term climate change compared to older types like R-22 or R-410A.

Incorporating renewable energy sources like solar or geothermal power enhances the sustainability of HVAC systems. Solar panels can be integrated into HVAC systems to supplement electricity needs, especially during peak sunlight hours. This reduces reliance on non-renewable energy sources, lowering overall energy costs.

Geothermal systems, on the other hand, use the stable temperatures underground to heat and cool spaces. These systems are incredibly efficient, as they require less energy to move heat rather than generate it.

If you're considering these green upgrades, here's how you can proceed:

- First, assess your local climate and the feasibility of installing solar panels. This involves looking at average sunlight hours and roof orientation.

- Consult with a professional to get a detailed analysis and quote for a solar-integrated HVAC system.

- If geothermal heating and cooling interest you, conduct a soil test to determine if your property can support geothermal wells.

- Once the initial assessments are complete, hire certified installers who have experience with these systems to ensure optimal performance and longevity.

- Make sure to apply for any available government incentives or rebates for installing renewable energy systems. These financial perks can offset some upfront costs, making such installations more accessible.

Taking these steps can pave the way for profoundly impactful energy savings and environmental benefits, enhancing the overall sustainability of your HVAC system.

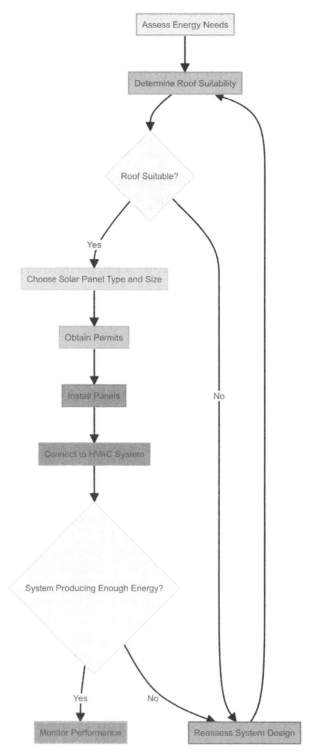

Integrating Solar Panels with Your HVAC
System: A Step-by-Step Guide

When it comes to HVAC design, sustainability is key. It's all about considering the energy use, emissions, and resource conservation throughout the entire lifecycle to ensure eco-friendly operation. Considering the complete lifecycle of your HVAC systems is crucial. This includes everything from manufacturing and installation to maintenance, repair, and eventual disposal.

When selecting HVAC systems, it's important to keep in mind options that are made from recycled materials and are designed for easy disassembly and recycling. By implementing this approach, we can reduce the resources taken from our planet and minimize the waste that accumulates in landfills when the product reaches the end of its life.

Additionally, adopting systems that use advanced filtration and air purification methods can improve indoor air quality, benefiting both human health and building longevity. Systems equipped with HEPA filters or ultraviolet germicidal irradiation (UVGI) can make significant strides toward maintaining a healthier indoor environment. These methods reduce pollutants and allergens that might otherwise be recirculated through traditional HVAC systems.

Paying attention to these design principles ensures that your HVAC system isn't just efficient but also minimally impactful over its entire lifespan. By investing in quality, sustainability-focused components, you contribute to a broader effort in environmental stewardship and social responsibility.

Key takeaways from green HVAC practices emphasize prioritizing energy efficiency, selecting eco-friendly refrigerants, integrating renewable energy sources, and adhering to sustainable design principles. These measures collectively forge a path toward reduced environmental footprints and improved human welfare, echoing a broader societal shift towards sustainability.

Studies show that buildings implementing high-efficiency HVAC systems report considerable reductions in energy consumption and operating costs (Green Heating and Cooling: Sustainable, Comfortable, Efficient - BlocPower, n.d.). According to Asim et al. (2022), incorporating renewable technologies can directly reduce greenhouse gas emissions and improve overall system performance, highlighting the tangible benefits of green HVAC practices.

Understanding and applying these concepts may feel daunting initially, but with careful planning and commitment, anyone can take meaningful steps towards a greener, more sustainable future.

Innovative HVAC Materials and Construction Methods

Materials have come a long way, allowing us to create HVAC components that are not only high-performance and lightweight, but also environmentally friendly. Imagine this: what if your home's heating and cooling system wasn't just effective but also considerably lighter and more sustainable?

Today's innovations in HVAC materials are setting us on a path where that vision is becoming a reality. Traditional HVAC systems, while reliable, often involve heavy and resource-intensive materials like metals for ductwork and extensive insulation layers.

However, thanks to advancements in material science, we're now seeing a shift toward more advanced, efficient alternatives.

For example, new composite materials are increasingly being used to build ductwork. These composites are not only lighter but also offer excellent durability and flexibility, which means easier installation and maintenance—less weighty metal and more efficient circulation paths for air.

Environmentally friendly materials are making their way into insulation and other HVAC components as well. This transition isn't just about reducing the carbon footprint; it also about enhancing the overall performance and lifespan of the systems we rely on every day.

Aerogels, known for their extreme lightness and superior thermal insulating properties, are an excellent example of these innovative materials. They provide incredible insulation with significantly less material, making HVAC units more compact and easier to install and manage.

New construction methods are being developed to improve the durability, energy efficiency, and ease of maintenance in HVAC installations. Constructing and maintaining our homes' systems is crucial for futureproofing them. Installing an HVAC system has become much easier and less invasive compared to the past.

New methods are revolutionizing this field, with a focus on enhancing system durability, energy efficiency, and ease of maintenance:

- Use modular construction techniques that allow for components to be easily replaced or upgraded, minimizing downtime and extending the system's life.

- Ensure that all connections and interfaces are standardized to streamline both installation and maintenance tasks.

- Incorporate smart diagnostics tools within the system to facilitate proactive maintenance, identifying potential issues before they become serious problems.

- Use eco-friendly and corrosion-resistant materials to enhance the longevity of the system even under harsh environmental conditions.

These guidelines aren't merely theoretical; they're being implemented in cutting-edge projects today with tremendous success.

Modular HVAC systems allow for quick replacement of faulty parts without needing a complete overhaul, thus saving time and resources. The integration of smart technologies, such as sensors and IoT devices, allows homeowners to monitor their systems in real-time, catching issues early and reducing the need for costly repairs.

Innovative materials like aerogels for insulation or composite materials for ductwork offer superior thermal performance and longevity. Consider how aerogels, despite being incredibly light, provide remarkable insulating properties.

What sets them apart from traditional materials is their unique structure that traps air within a solid form, drastically reducing heat transfer. This makes them perfect for applications where space is at a premium yet high thermal resistance is crucial.

Such materials help in constructing HVAC components that are not only efficient in terms of thermal management but also robust and long-lasting.

Composite materials are becoming increasingly popular because of their versatility and strength, just like aerogels. Composites are specifically engineered to endure tough conditions and deliver reliable performance for long durations, unlike conventional materials that may corrode or deteriorate over time. By utilizing these advanced materials, HVAC systems can be more efficient and durable, resulting in fewer replacements and less maintenance. This not only saves costs in the long run but also helps minimize the environmental impact.

In the future, HVAC systems could potentially integrate cutting-edge nanotechnology or smart materials to enhance their performance and promote sustainability. The future of HVAC technology is filled with immense promise, particularly due to exciting advancements in nanotechnology and smart materials.

Picture this: an HVAC system that can adapt its thermal properties in real-time, responding to the surrounding environment. It may sound like something out of a sci-fi movie, but believe it or not, this technology is not as far off as we might think.

Nanotechnology could revolutionize how we manage temperature and airflow within our homes. Nano-coatings on HVAC components could dramatically improve efficiency by reducing friction and wear, leading to quieter and longer-lasting systems. Additionally, nanoscale materials could enhance heat exchange processes, making systems more responsive and energy efficient.

Smart materials, on the other hand, bring adaptability to the table. These materials can respond to changes in temperature or humidity automatically. For instance, shape-memory alloys or polymers might adjust the flow of air through ducts or vents based on real-time data, optimizing comfort and energy use. By leveraging these innovations, future HVAC systems will be smarter, more autonomous, and better aligned with our sustainability goals.

As we navigate the intricacies of the current climate crisis and persistent energy demands, it's apparent that more efficient, durable, and environmentally friendly HVAC systems are not just desirable—they're essential.The shift towards sustainable, high-performance HVAC systems represents a significant step forward in creating homes that are comfortable, cost-effective, and kind to our environment.

Ultimately, the future of HVAC technology is about creating environments where people can thrive comfortably and sustainably. The marriage of human expertise and cutting-edge technology promises resilient and adaptive solutions.

References

jenks2026. (2024). The Future of Building Automation: Insights from Industry Leaders. Green.org. Retrieved from https://green.org/2024/01/30/the-future-of-building-automation-insights-from-industry-leaders/

BlocPower. (No date). Green Heating and Cooling: Sustainable, Comfortable, Efficient. BlocPower. Retrieved from https://blocpower.org/posts/green-hvac-technology

MITCHELL, B. (2023). The HVAC/R Industry vs. Automation. Northeast Technical Institute. Retrieved from https://ntinow.edu/hvac-industry-vs-automation/

HVAC School. (2017). Advancements in HVAC-R Technology. HVACSCHOOL.ORG. https://www.hvacschool.org/hvac-technology/

Asim, N., Badiei, M., Mohammad, M., Razali, H., Rajabi, A., Haw, L. C., & Ghazali, M. J. (2022). Sustainability of Heating, Ventilation and Air-Conditioning (HVAC) Systems in Buildings—An Overview. International Journal of Environmental Research and Public Health, 19(2), 16. https://doi.org/10.3390/ijerph19021016

Lindner, J. (2024). AI In The Hvac Industry Statistics. Gitnux. Retrieved from https://gitnux.org/ai-in-the-hvac-industry/

Green, J. (2024). The future of building automation: Insights from industry leaders. Green.org. Retrieved from https://green.org/2024/01/30/the-future-of-building-automation-insights-from-industry-leaders/

Leffer, L. (n.d.). New Air-Conditioning Technology Could Be the Future of Cool. Scientific American. https://www.scientificamerican.com/article/new-air-conditioning-technology-could-be-the-future-of-cool1/

Asim, N., Badiei, M., Mohammad, M., Razali, H., Rajabi, A., Lim, C. H., ... Ghazali, M. J. (2022). Sustainability of Heating, Ventilation and Air-Conditioning (HVAC) Systems in Buildings—An Overview. International Journal of Environmental Research and Public Health, 19(2), 1-16. https://doi.org/10.3390/ijerph19021016

Practical Application

Maintaining an HVAC system is more than just a routine task; it's a commitment to ensuring comfort, efficiency, and longevity in your home or workspace. Many homeowners face common yet preventable problems with their HVAC systems due to a lack of structured maintenance plans.

For example, dirty air filters can force your system to work harder, leading to higher energy costs and potential breakdowns. Similarly, failing to routinely inspect electrical connections can result in unsafe operations or even complete system failures.

These issues often culminate during extreme weather conditions, when the demand for heating or cooling is highest, leaving you vulnerable to expensive emergency repairs and uncomfortable living conditions. By establishing a regular maintenance schedule and using simple diagnostic tools like multimeters and digital thermometers, these challenges can be effectively mitigated.

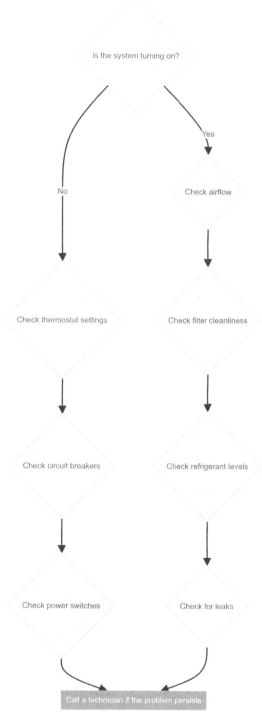

Troubleshooting Common HVAC Issues: A Step-by-Step Guide

The ultimate goal is that you are equipped with actionable insights to maintain your HVAC system efficiently, contributing to its reliability, performance, and your overall peace of mind.

Developing a Personalized HVAC Maintenance Plan

Adopting a regular maintenance schedule is paramount to avoid unexpected breakdowns and ensure optimal performance of your HVAC system. Regular inspections help identify and address issues early, making the whole process more manageable and less costly in the long run.

The first step toward developing a comprehensive maintenance plan is establishing a regular maintenance schedule. This includes inspecting and servicing key HVAC components like filters, coils, and thermostats. An annual pre-season check-up from a professional contractor can significantly enhance the system's efficiency.

Spring and fall are ideal times for these check-ups, as highlighted by ENERGY STAR® (n.d.). Pre-season checks prevent last-minute rushes during peak seasons and ensure your system is prepared when you need it most.

To make this practical, let's create a checklist of tasks. Here is what you can do to establish your schedule:

- Schedule annual check-ups with a certified HVAC contractor in spring for cooling systems and fall for heating systems.

- Inspect and clean or replace air filters monthly. Dirty filters force your system to work harder, increasing energy costs and the risk of system failure.

- Examine the thermostat settings regularly to ensure they maintain the desired temperature efficiently.

Using maintenance checklists and tools systematically will also contribute to the reliability and longevity of your HVAC system. A well-crafted checklist serves as a guide, helping you cover all necessary aspects of the system's maintenance. ENERGY STAR® provides a robust checklist that outlines essential tasks (ENERGY STAR®, n.d.). Following these guidelines ensures nothing is overlooked.

- Use tools such as a digital thermometer and a voltage meter to assess system components accurately.

- Start by tightening all electrical connections and measuring voltage on motors to prevent unsafe operations and prolong component life.

- Lubricate moving parts to reduce friction and improve efficiency.

- Clear the condensate drain to avoid water damage and regulate indoor humidity levels effectively.

Incorporating energy-saving practices into your maintenance routine has multiple benefits. It not only reduces operating costs but also contributes to broader environmental objectives. Two effective methods to achieve this are adjusting thermostat settings and sealing ductwork.

Sealing ducts is particularly vital, as around 20-30% of heated or cooled air can be lost through leaks, holes, and poorly connected ducts in typical homes (ENERGY STAR®, n.d.).

For optimal energy savings:

- Install a smart thermostat to optimize temperature settings based on real-time needs. This simple upgrade can significantly cut down on wasted energy, offering convenience and cost savings.

- Use weather stripping or caulk to seal ducts in accessible areas such as attics, basements, and crawl spaces. Avoid using traditional duct tape, which lacks longevity.

- Ensure proper insulation of ducts to maintain the desired temperature of air being transported throughout your home.

Documenting maintenance tasks and observations plays an integral role in tracking HVAC system performance and making informed decisions for future improvements. Keeping meticulous records empowers homeowners and technicians to spot trends or recurrent issues, facilitating proactive rather than reactive maintenance.

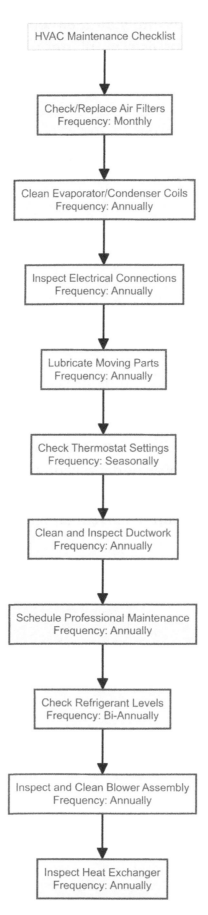

HVAC Maintenance Log

To document effectively:

- Use a dedicated maintenance journal or digital tool to log every task, noting any performed repairs or replacements.

- Include dates, specific actions taken, and observations about system performance.

- Review recorded data periodically to identify patterns or emerging issues before they escalate into significant problems.

When you make it a habit to document things, you get better at foreseeing possible problems and can plan maintenance in advance. This helps you avoid expensive last-minute repairs.

Consistent maintenance isn't just about preventing mechanical failures; it plays a critical role in sustaining comfort and reducing overall operating costs. A system that operates efficiently not only lasts longer but also maintains a comfortable environment without spiking utility bills.

For instance, a clean filter can reduce energy consumption by up to 15%, while sealing ducts can resolve significant inefficiencies, ensuring every room in your home remains comfortable year-round (ENERGY STAR®, n.d.)

While personal responsibility in maintaining your HVAC system is essential, there are times when seeking professional assistance becomes necessary. Contractors bring expertise and tools that might be beyond the reach of average homeowner. Also, some tasks, like handling refrigerants or performing complex electrical diagnostics, require certification and specialized knowledge.

Here's how to have a structured and diligent approach when it comes to HVAC maintenance:

- Establish a regular maintenance schedule with professional check-ups to ensure your system remains in top condition.

- Use comprehensive checklists and appropriate tools for systematic identification and resolution of potential issues.

- Incorporate energy-saving practices to help reduce operational costs while promoting sustainable living.

- Document maintenance activities to provide valuable insights for continuous improvement and proactive problem-solving.

Remember, consistent maintenance is not merely about keeping your HVAC system running; it's about optimizing its performance to maximize comfort, efficiency, and lifespan. Simple, regular care can lead to significant long-term benefits, enhancing both the system's reliability and your peace of mind.

By following these guidelines, you are investing in a more comfortable, efficient, and cost-effective future for your home.

Effective Troubleshooting Techniques

When it comes to effective troubleshooting of HVAC systems, being familiar with common issues, employing diagnostic tools correctly, adhering to safety protocols, and continuously upgrading your skills are crucial steps.

Familiarity with frequent problems such as airflow restrictions, refrigerant leaks, and electrical component failures enables you to diagnose issues more efficiently. Airflow restrictions can be caused by dirty air filters, blocked ducts, or improperly sized ductwork. Regularly changing air filters every three months, or more frequently if you have pets, can prevent these problems (Academy, 2021).

Changing Your Air Filter: A Simple DIY Maintenance Task

Refrigerant leaks not only affect cooling efficiency but also pose environmental concerns. Spotting signs like ice buildup on the evaporator coils or a noticeable decrease in cooling performance can indicate a leak. Electrical issues usually manifest as tripped circuit breakers or faulty components like capacitors or contactors that need regular inspection and replacement to ensure the HVAC system runs smoothly.

Using the right diagnostic tools is paramount. A multimeter, pressure gauges, and leak detectors are indispensable in identifying the root cause of HVAC malfunctions. To perform an accurate diagnosis:

- Use a multimeter to check voltage, current, and resistance. It is essential for diagnosing electrical issues, such as ensuring there is no short circuit that could trip the breaker.

- Pressure gauges help measure the system's refrigerant levels, confirming whether there is a refrigerant leak. Each type of refrigerant operates at specific pressures, so matching those values is vital.

- Leak detectors pinpoint the exact location of refrigerant leaks, allowing for targeted repairs.

Equip yourself with a set of manifold gauges for testing system pressures. These gauges give you a snapshot of how well the refrigerant system is operating. Temperature probes can assist in measuring the superheat and subcooling, providing further insights into the system's functionality.

Safety protocols cannot be overemphasized. Working with HVAC systems involves risks like electric shocks, refrigerant exposure, and physical injuries from moving parts.

Always follow these safety guidelines:

- Wear appropriate personal protective equipment (PPE) including gloves, safety glasses, and masks when necessary to protect against harmful substances.

- Before beginning any work, ensure the power supply is turned off and verified with a multimeter to avoid electric shocks.

- Never bypass safety mechanisms while diagnosing issues. For example, don't remove or tamper with covers or sensors unless absolutely necessary and if you do, make sure the unit is powered off.

- Follow proper handling and disposal procedures for refrigerants to comply with environmental and safety regulations (ENERGY STAR®, n.d.).

Regular training and hands-on experience significantly enhance your troubleshooting skills. Enrolling in training programs, attending workshops, and getting certified through reputable institutions can keep you up to date with the latest advancements and industry standards.

Most importantly, on-the-job experience refines your ability to tackle diverse HVAC challenges effectively.

Here are some steps to upgrade your troubleshooting skills:

- Participate in accredited HVAC technician programs to understand new technologies and methodologies.

- Attend industry workshops and seminars to learn from experienced professionals.

- Take on small projects and gradually increase complexity as your confidence grows.

- Stay updated with HVAC literature and online resources for continuous learning and improvement.

By honing these skills, you ensure that your troubleshooting approach evolves with the industry standards, making you proficient in resolving even the most intricate HVAC issues.

Troubleshoot HVAC systems effectively to avoid expensive repairs and minimize downtime. This way, you can ensure that the systems are running at their best and maintaining a comfortable indoor environment.

Having a good grasp of airflow dynamics, refrigerant flow, and electrical functioning is crucial when it comes to troubleshooting. Using diagnostic tools correctly allows us to pinpoint issues with accuracy, while following strict safety protocols guarantees the well-being of both individuals and the system.

Furthermore, constantly improving your skills through education and practice strengthens your ability to solve a wide range of challenging HVAC issues.

Engaging in Hands-On HVAC Projects

Engaging in hands-on HVAC projects is an invaluable way to enhance your skills and deepen your understanding of HVAC principles.

Let's dive into some practical ways you can build expertise and make meaningful contributions to your community, all while preparing for a successful career in this high-demand field.

One of the best starting points is to undertake small DIY projects. Simple yet impactful tasks like installing programmable thermostats or improving insulation can provide firsthand experience with HVAC system modifications. These projects allow you to familiarize yourself with basic tools and techniques, as well as understand the intricacies of how HVAC systems function.

- Begin by researching the type of thermostat that best fits your needs. Programmable thermostats come in various models with different capabilities.

- Once you've acquired your thermostat, read its installation manual thoroughly. Each model may have unique steps or considerations.

- Turn off the power supply to your HVAC system before beginning any work. Safety is paramount.

- Carefully remove the old thermostat, making sure to disconnect wiring systematically. Take note of wire labels or capture a photo for reference.

- Follow the installation steps detailed in the manual to install your new thermostat. Secure connections are crucial to ensure proper functionality.

- After installation, test the thermostat to confirm it's working correctly. Make adjustments as necessary to optimize your settings for energy efficiency.

Another fantastic avenue for hands-on learning is volunteering for community HVAC initiatives or apprenticeships. By collaborating with industry professionals, you gain valuable insights into real-world HVAC applications. Volunteering not only enhances your technical skills but also fosters a spirit of community and social responsibility.

Sign up for HVAC workshops, seminars, and networking events to stay abreast of industry trends. These events provide opportunities to connect with experienced professionals who can offer mentorship. Regularly attending these gatherings ensures you remain up to date with technological advancements and best practices within the HVAC industry.

Here's what you need to do to make the most of these opportunities:

- Look up local HVAC workshops or trade shows scheduled in your area and register in advance.

- Participate actively in seminars and ask questions to clarify doubts or expand your knowledge.

- Network with fellow attendees and presenters. Exchanging contact information can open doors to future collaborations or mentorship.

- Follow up on connections made during these events through LinkedIn or email to solidify professional relationships.

Documenting your project outcomes, lessons learned, and successes is another crucial step. Building a portfolio showcasing your HVAC skills and accomplishments prepares you for future career opportunities.

This portfolio serves as tangible evidence of your capabilities and dedication, setting you apart in job applications or interviews.

Here is how you can document effectively:

- Keep a detailed log of each project you undertake, describing the objectives, challenges faced, solutions implemented, and results achieved.

- Capture photographs or videos of your work at various stages. Visual documentation adds credibility and makes your portfolio more engaging.

- Reflect on the lessons learned from each project. Highlight what went well and areas where there was room for improvement. This reflection demonstrates growth and adaptability.

- Organize your portfolio in a clear, professional format. Include a table of contents, section dividers, and concise summaries for easy navigation.

Hands-on projects serve as powerful experiential learning opportunities. They allow you to refine your HVAC skill set and build a foundation for career advancement.

With the demand for HVAC technicians expected to grow significantly (Sweeney, 2023), taking proactive steps now can ensure you're well-prepared to meet industry needs. Engaging in practical projects, volunteering, participating in educational events, and documenting your achievements are all strategies that will pay dividends in your professional journey.

By focusing on both personal development and contributing to broader community efforts, you're not just advancing your own career—you're helping to address a critical shortage in skilled HVAC professionals. The importance of getting involved cannot be overstated; it's about building expertise, ensuring safety and efficiency, and enriching communities.

Also, think about staying in touch with people in the HVAC industry through professional groups, online forums, and social media. Regular participation in these spaces keeps you up to date on the latest changes in your field and lets you get help from seasoned professionals.

It's also worth noting the diversity of job roles within the HVAC industry, from technicians and installers to engineers and managers (Sweeney, 2023). Exploring different specializations can help you identify where your interests and strengths lie.

So, roll up your sleeves and start exploring the world of HVAC hands-on! Engage with your community, learn from seasoned professionals, and never stop honing your skills. Your dedication today will unlock tomorrow's opportunities in this vital and rewarding industry.

Continuous Education and Professional Growth

In the dynamic world of HVAC, continuous education and professional growth are paramount to staying relevant and enhancing one's career prospects. Let's dive into some essential avenues for development.

Pursuing Industry Certifications: Obtaining industry certifications like NATE (North American Technician Excellence) or HVAC Excellence can significantly bolster your credibility and expertise in the HVAC field. These credentials not only validate your knowledge but also open up new career opportunities. Here's a practical approach to earn them:

- Begin by researching which certification aligns best with your career goals.

- Enroll in preparatory courses, either online or at local institutions, focused on the requisite skills and knowledge areas.

- Dedicate regular study time each week to review materials and practice hands-on applications.

- Register for the exam, ensuring you meet any prerequisites such as work experience or education requirements.

- Finally, take the exam when you feel fully prepared, and don't hesitate to use practice tests to gauge your readiness.

Attending Seminars, Webinars, and Trade Shows: The HVAC industry is perpetually evolving with advancements in technology, changes in regulations, and emerging best practices. It is important to keep up with these changes if you want to stay competitive. Here's how to get the most out of learning events:

- Regularly check industry websites, professional organizations, and trade publications for information about upcoming seminars, webinars, and trade shows.

- Register for events that focus on topics relevant to your practice area or those that will help broaden your technical skills and industry knowledge.

- Participate actively by taking notes, asking questions, and networking with speakers and fellow attendees.

- After attending, review your notes and integrate the newfound insights into your daily work routines or projects.

Seeking Mentorship and Joining Professional Organizations: Interaction with seasoned professionals and being part of industry groups like ASHRAE (American Society of Heating, Refrigerating and Air-Conditioning Engineers) can provide invaluable guidance and resources. For those looking to connect with mentors and expand their network:

- Identify experienced professionals within your company or through industry events who have a track record of success in areas of interest to you.

- Approach potential mentors respectfully, expressing your admiration for their work and your desire to learn from their experiences.

- Join professional organizations like ASHRAE to access a wealth of resources, including technical literature, standards, and professional development opportunities.

- Participate in organization meetings, forums, and volunteer activities to build relationships and gain insights from a broader community of HVAC professionals.

Continuously Upgrading Technical Skills and Knowledge: Online courses, workshops, and on-the-job training are excellent avenues for keeping abreast of industry trends and enhancing your technical acumen. Continuous learning keeps you adaptable and relevant. Here's how to structure your ongoing education:

- Assess your current skill set and identify gaps or areas where you want to improve or expand your expertise.

- Research and enroll in online courses or workshops that cover these areas, ensuring they offer practical, hands-on components where possible.

- Balance theoretical learning with on-the-job practice, applying what you've learned in real-world scenarios under the guidance of more experienced colleagues if necessary.

- Stay updated with new releases of HVAC standards and codes, and consider cross-training in related fields such as refrigeration, energy management, and building automation.

Through these steps, lifelong learning becomes second nature, compelling professionals to stay ahead in a competitive and innovative environment.

By dedicating yourself to continuous improvement, you'll not only enhance your own career prospects but also contribute to the advancement of the entire HVAC industry. Let's aim for a future where every HVAC professional is empowered with the knowledge, skills, and resources needed to excel and innovate.

References

Institute of Heating & Air. (2024). Maximize Your HVAC/R/SM Business Growth with Our Trade Shows. Blog. Retrieved from https://www.ihaci.org/maximize-business-growth-with-trade-shows/

ASHRAE. (n.d.). Resources. https://www.ashrae.org/communities/young-engineers-in-ashrae-yea/resources

Energy.gov. (n.d.). Preventative maintenance for commercial HVAC equipment. Retrieved from https://betterbuildingssolutioncenter.energy.gov/solutions-at-a-glance/preventative-maintenance-commercial-hvac-equipment

Hvac Global. (2022). HVAC Trouble Shooting Guide. Retrieved from https://hvacglobal.org/hvac-trouble-shooting-guide/

ENERGY STAR®. (n.d.). Maintenance Checklist. Retrieved from https://www.energystar.gov/saveathome/heating-cooling/maintenance-checklist

Fortis Colleges & Institutes. (n.d.). How Tech is Changing HVAC Jobs. https://www.fortis.edu/blog/skilled-trades/how-tech-is-changing-hvac-jobs.html

Sweeney, A. (2023). Why HVAC/R Technicians are in High Demand. Northeast Technical Institute. Retrieved from https://ntinow.edu/why-hvac-r-technicians-are-in-high-demand/

Education Directory. (n.d.). AEE Education: Filter By Certification Trainings. Retrieved from https://education.aeecenter.org/certification-trainings

ENERGY STAR®. (n.d.). How to Keep Your HVAC System Working Efficiently. Retrieved from https://www.energystar.gov/products/ask-the-experts/how-keep-your-hvac-system-working-efficiently

ASHRAE. (n.d.). Learn more about Supplier Webinars at ashrae.org. Retrieved from https://www.ashrae.org/technical-resources/supplier-provided-learning/supplier-webinars

Florida Academy. (2021). DIY HVAC Repair Guide: How to Fix Your HVAC System. Florida-Academy. Retrieved from https://florida-academy.edu/diy-hvac-repair-guide/

Golden Rules of HVAC

Why This Chapter Is Here

The 'Golden Rules of HVAC' chapter is designed to serve as a quick reference guide, summarizing the most critical pieces of learning from the entire book. Inspired by Pareto's Principle, this section highlights the 20% of knowledge that will cover 80% of HVAC needs. Whether you're on the job and need a quick refresher or simply want to review the most essential points, this chapter is your go-to cheat sheet.

HVAC System Fundamentals

Understand the Basics: HVAC stands for Heating, Ventilation, and Air Conditioning. These systems regulate indoor climate and ensure comfort. They are essential in residential, commercial, and industrial buildings to maintain optimal temperature, humidity, and air quality.

Key Components: The main parts of an HVAC system include:

- Furnace: Heats the air in the system.
- Heat Exchanger: Transfers heat from the combustion process to the air.
- Evaporator Coil: Cools the air by evaporating the refrigerant.
- Condenser Coil: Releases heat from the refrigerant outside.
- Refrigerant Lines: Carry the refrigerant between components.
- Thermostat: Controls the system and maintains desired temperature.
- Ductwork: Distributes conditioned air throughout the building.

Refrigeration Cycle

Four Main Components: The refrigeration cycle involves four key components: the evaporator, the compressor, the condenser, and the expansion valve.

- Evaporator: Absorbs heat from the indoor air, causing the refrigerant to evaporate and cool the air.

- Compressor: Increases the pressure of the refrigerant, raising its temperature.

- Condenser: Releases the absorbed heat to the outside air, causing the refrigerant to condense back into a liquid.

- Expansion Valve: Lowers the pressure of the refrigerant, causing it to cool rapidly before it returns to the evaporator.

Cycle Explanation: The refrigerant absorbs heat in the evaporator, is compressed to a high-pressure gas by the compressor, releases heat in the condenser, and is expanded and cooled by the expansion valve before returning to the evaporator.

Efficiency Tips: Ensure the system is properly charged with refrigerant and that all components are clean and free of obstructions for optimal efficiency.

Installation Essentials

Proper Sizing: Ensure the system is appropriately sized for the space. An undersized unit will struggle to maintain comfort, while an oversized unit will cycle on and off frequently, leading to inefficiency and increased wear. Conduct a load calculation using the Manual J method to determine the correct size.

Quality Insulation: Proper insulation of ductwork and sealing of joints prevent energy loss and improve efficiency. Use high-quality insulation materials and ensure all seams are tightly sealed.

Accurate Leveling: Ensure units are level to prevent operational issues and extend the lifespan of components. An uneven installation can cause stress on parts and lead to premature failure. Use a level during installation to verify accuracy.

Airflow Management: Ensure that airflow is unobstructed. Properly size and install ductwork to minimize resistance and maximize efficiency. Use dampers to balance airflow and ensure even distribution throughout the space.

Maintenance Must-Dos

Regular Filter Changes: Replace filters every 1-3 months to maintain air quality and system efficiency. Clogged filters restrict airflow and can cause the system to overheat or freeze. Use filters with the appropriate MERV rating for your system.

Scheduled Inspections: Perform bi-annual inspections, ideally in the spring and fall, to catch and fix potential issues early. Check all components for wear and tear, clean as necessary, and verify system performance.

Clean Coils: Regularly clean the evaporator and condenser coils to maintain optimal heat exchange efficiency. Dirty coils reduce the system's ability to transfer heat and can lead to increased energy consumption. Use a soft brush and a coil cleaner to remove debris.

Check Refrigerant Levels: Ensure refrigerant levels are within the manufacturer's specifications. Low levels can indicate leaks and lead to reduced efficiency and system damage. If levels are low, locate and repair any leaks before recharging the system.

Inspect Electrical Connections: Tighten all electrical connections and check for signs of wear or corrosion. Faulty connections can cause system failures and pose a safety hazard. Use a multimeter to verify voltage and current readings.

Troubleshooting Tips

Check Thermostat Settings: Ensure the thermostat is set correctly and functioning properly before diagnosing more complex issues. Verify the programming and temperature settings, and replace batteries if needed.

Inspect Airflow: Check for obstructions in vents and ductwork that could impede airflow and reduce efficiency. Ensure all registers are open and unobstructed. Clean or replace filters as necessary.

Listen for Unusual Noises: Strange noises often indicate mechanical problems. Identify and address these issues promptly. Common noises include:

- Rattling: Loose components or debris in the system.

- Hissing: Refrigerant leaks or ductwork issues.

- Squealing: Worn belts or motor bearings.

- Banging: Loose or broken parts.

Examine System Cycling: If the system is cycling on and off frequently, check for issues such as incorrect thermostat settings, dirty filters, or improper refrigerant levels. Frequent cycling can also indicate an oversized unit.

Energy Efficiency Hacks

Programmable Thermostat: Use a programmable thermostat to optimize heating and cooling schedules and reduce energy consumption. Set temperatures lower when the building is unoccupied and higher when it is in use. Smart thermostats can learn patterns and adjust settings automatically.

Seal Leaks: Ensure all ducts are sealed properly to prevent energy loss. Use mastic sealant or metal tape for the best results. Leaky ducts can account for significant energy losses, especially in unconditioned spaces like attics or basements.

Upgrade Insulation: Improve home insulation to reduce the load on HVAC systems and save on energy costs. Insulate attics, walls, and floors to maintain desired temperatures with less energy. Consider adding radiant barriers in attics to reflect heat away from the living space.

Use Ceiling Fans: Ceiling fans can help distribute air more evenly and reduce the load on HVAC systems. In summer, run fans counterclockwise to create a cooling breeze. In winter, run them clockwise to push warm air down from the ceiling.

Optimize Windows: Use energy-efficient windows and coverings to reduce heat loss and gain. Install double-pane windows and use thermal curtains or blinds to insulate against temperature extremes.

Safety Precautions

Turn Off Power: Always shut off power to the HVAC unit before performing any maintenance or repairs to prevent electrical shocks. Use the circuit breaker or a disconnect switch to ensure the system is completely de-energized.

Proper Ventilation: Ensure adequate ventilation when working with refrigerants and other chemicals to avoid inhalation hazards. Work in well-ventilated areas and use fans or exhaust systems to disperse fumes.

Use Proper PPE: Wear personal protective equipment such as gloves, safety glasses, and masks to protect against injury and exposure to harmful substances. Follow safety guidelines and manufacturer recommendations for handling chemicals and working with electrical components.

Avoid Asbestos: Be cautious when working in older buildings that may contain asbestos. If asbestos is suspected, do not disturb it and seek professional remediation. Asbestos fibers can cause serious respiratory issues if inhaled.

Regulatory Compliance

Stay Updated: Keep abreast of the latest HVAC codes and regulations to ensure all work meets legal standards. Regularly review updates from organizations such as ASHRAE, EPA, and local building codes.

Documentation: Maintain thorough records of installations, maintenance, and repairs for compliance and warranty purposes. Detailed records help track system performance and identify recurring issues.

Certification: Ensure all technicians are properly certified and trained according to industry standards. Certification demonstrates competency and commitment to quality work. Encourage ongoing education and training to stay current with industry advancements.

Tools and Equipment

Essential Tools: List of must-have tools for HVAC professionals, such as multimeters, gauges, vacuum pumps, etc.

- Multimeters: Used for measuring voltage, current, and resistance.
- Gauges: Essential for measuring refrigerant pressure.
- Vacuum Pumps: Used to remove air and moisture from the system.

Maintenance Tools: Specific tools needed for regular maintenance tasks.

- Coil Cleaner: Used to clean the evaporator and condenser coils.
- Fin Comb: Used to straighten bent fins on coils.
- Refrigerant Scale: Used to measure refrigerant accurately.

Customer Service Best Practices

Effective Communication: Tips on how to communicate clearly and effectively with customers.

- Listen actively to customer concerns and questions.

- Provide clear explanations and avoid technical jargon.
- Follow up with customers to ensure satisfaction.

Managing Expectations: Strategies for setting and managing customer expectations regarding service and repair timelines.

- Provide realistic timelines for service and repairs.
- Keep customers informed of any delays or changes.
- Be transparent about costs and potential additional charges.

Advanced Troubleshooting Techniques

Detailed Diagnostics: Step-by-step guide for diagnosing more complex HVAC issues.

- Use diagnostic tools to pinpoint issues accurately.
- Check for error codes and refer to the manufacturer's manual.
- Perform a visual inspection for any obvious signs of damage.

Common Problems and Solutions: A list of frequent HVAC problems and their solutions.

- No Cooling: Check thermostat settings and refrigerant levels.
- Strange Noises: Inspect for loose components or debris.
- Poor Airflow: Clean or replace filters and check for duct obstructions.

Environmental Considerations

Eco-friendly Practices: Tips for implementing environmentally friendly HVAC practices.

- Use energy-efficient equipment and materials.
- Encourage regular maintenance to reduce energy consumption.
- Implement recycling programs for old equipment.

Refrigerant Management: Best practices for handling and disposing of refrigerants to minimize environmental impact.

- Follow EPA guidelines for refrigerant handling and disposal.

- Use reclaim and recycling equipment for refrigerants.
- Properly label and store refrigerant containers.

Conclusion

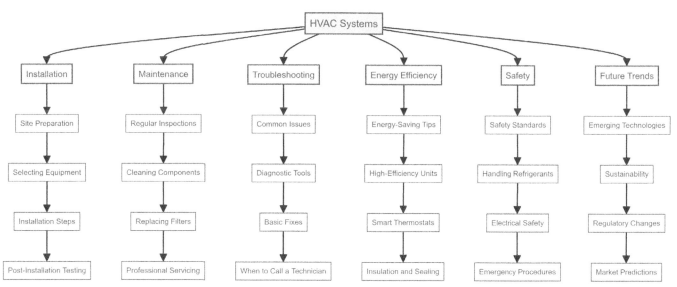

Key Takeaways from HVAC for Beginners: Your Roadmap to a Comfortable and Efficient Home

As we conclude our key learnings and practical applications of HVAC systems, it's evident that understanding both theoretical concepts and their real-world applications is crucial for anyone involved with these systems.

Throughout our discussion, we emphasized how mastering thermodynamics, fluid mechanics, and heat transfer is foundational. These principles are not mere abstracts but practical tools aiding homeowners, DIY enthusiasts, and novice technicians alike in maintaining and optimizing HVAC systems.

The journey began with fundamental concepts, highlighting how knowledge of energy transfer and airflow can significantly impact system efficiency and comfort. Installing systems correctly according to manufacturer specifications prevents unnecessary energy consumption and ensures safety. Regular maintenance, such as cleaning filters and checking for leaks, extends the lifespan of these systems while reducing costs.

Our focus also expanded to the socio-economic dimensions of HVAC systems. Individual actions in adopting energy-saving practices, like using programmable thermostats or opting for high-efficiency units, align with broader sustainability goals. The interplay between government regulations and corporate responsibility is pivotal, encouraging innovation and providing consumers with energy-efficient choices through incentives and clear standards.

In reflecting on these insights, it's clear that HVAC systems deeply intertwine with daily life, environmental stewardship, and economic considerations. Whether you are maintaining your home's comfort or seeking to advance professionally, the principles discussed here provide a strong foundation.

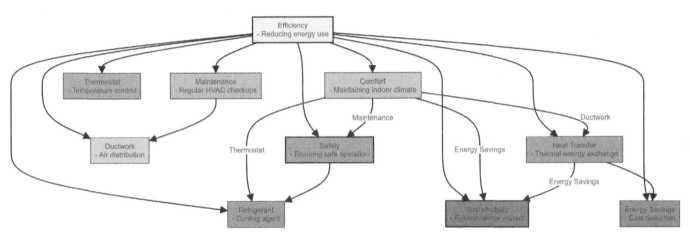

HVAC Essentials: A Visual Summary of Key Terms and Concepts

As we move forward, let's remember that continuous learning and application of best practices in HVAC systems hold the key to efficiency, health, and sustainability. Each step you take in understanding and managing these systems contributes to a more comfortable and environmentally friendly living space.

Here's to embracing the responsibility and reaping the benefits of well-maintained HVAC systems, knowing that our collective efforts pave the way towards a sustainable future.

Glossary

1. **AFUE (Annual Fuel Utilization Efficiency):** A measure of a furnace's efficiency, representing the percentage of fuel converted into heat.

2. **AHRI (Air-Conditioning, Heating, and Refrigeration Institute):** An organization that develops standards for HVAC equipment and certifies their performance.

3. **Air conditioner:** A system that removes heat from indoor air to cool and dehumidify a space.

4. **Air filter:** A device that removes dust, pollen, and other particles from the air passing through an HVAC system.

5. **Air handler:** The indoor unit of an HVAC system that houses the blower, evaporator coil, and filter.

6. **Airflow:** The movement of air through a duct system or space.

7. **ASHRAE (American Society of Heating, Refrigerating and Air-Conditioning Engineers):** A global organization that develops standards and guidelines for HVAC systems.

8. **BMS (Building Management System):** A computer-based system that controls and monitors a building's mechanical and electrical equipment, including HVAC systems.

9. **BTU (British Thermal Unit):** A unit of heat energy; the amount of heat needed to raise the temperature of one pound of water by one degree Fahrenheit.

10. **Blower:** A fan that moves air through an HVAC system.

11. **Building codes:** Regulations that govern the design, construction, and maintenance of buildings, including HVAC systems.

12. **Capacity:** The amount of heat an HVAC system can add to or remove from a space.

13. **Carbon footprint:** The total amount of greenhouse gases produced by an individual, organization, or product.

14. **Central air conditioner:** An HVAC system that cools an entire building from a single unit.

15. **Charge:** The amount of refrigerant in an HVAC system.

16. **Chilled water system:** A cooling system that uses chilled water circulated through pipes to cool air or equipment.

17. **Circuit breaker:** An electrical switch that automatically interrupts an electric circuit to prevent damage from overload or short circuits.

18. **Compressor:** A device that pressurizes refrigerant vapor in an HVAC system.

19. **Condensation:** The process by which a gas changes into a liquid.

20. **Condenser coil:** The outdoor coil of an air conditioner or heat pump that releases heat from the refrigerant to the outside air.

21. **Conduction:** The transfer of heat through direct contact between materials.

22. **Convection:** The transfer of heat through the movement of fluids (liquids or gases).

23. **Cooling load:** The amount of heat that needs to be removed from a space to maintain a desired temperature.

24. **Damper:** A valve or plate that regulates airflow in a duct system.

25. **Dehumidifier:** A device that removes moisture from the air.

26. **Ductless mini-split system:** An HVAC system that uses individual indoor units connected to an outdoor unit, without the need for ductwork.

27. **Ductwork:** A system of ducts that distribute conditioned air throughout a building.

28. **Economizer:** An HVAC control that uses outside air to cool a building when conditions are favorable.

29. **Efficiency:** The ratio of useful output to energy input in an HVAC system.

30. **Electric heating system:** A heating system that uses electricity to generate heat.

31. **Energy audit:** An assessment of a building's energy use to identify opportunities for

improvement.

32. **Energy recovery ventilator (ERV):** A ventilation system that transfers both heat and moisture between incoming and outgoing airstreams.

33. **Environmental Protection Agency (EPA):** A U.S. federal agency responsible for protecting human health and the environment.

34. **Evaporator coil:** The indoor coil of an air conditioner or heat pump that absorbs heat from the indoor air.

35. **Expansion valve:** A valve that controls the flow of refrigerant into the evaporator coil.

36. **Filter drier:** A device that removes moisture and debris from the refrigerant in an HVAC system.

37. **Forced air heating system:** A heating system that uses a furnace to heat air, which is then distributed throughout a building by a blower and ductwork.

38. **Furnace:** A device that burns fuel (gas, oil, or electricity) to generate heat.

39. **Geothermal heat pump:** An HVAC system that uses the earth's constant temperature to heat and cool buildings.

40. **Heat exchanger:** A device that transfers heat between two fluids without mixing them.

41. **Heat gain:** The amount of heat added to a space from sources such as sunlight, people, and equipment.

42. **Heat loss:** The amount of heat lost from a space to the outside environment.

43. **Heat pump:** An HVAC system that can both heat and cool a space by transferring heat between the indoor and outdoor environments.

44. **Heating load:** The amount of heat that needs to be added to a space to maintain a desired temperature.

45. **Heat pump:** An HVAC system that can both heat and cool a space by transferring heat between the indoor and outdoor environments.

46. **HEPA (High-Efficiency Particulate Air) filter:** A type of air filter that can remove at least 99.97% of particles that are 0.3 microns in size.

47. **Humidifier:** A device that adds moisture to the air.

48. **Humidity:** The amount of moisture in the air.

49. **HVAC (Heating, Ventilation, and Air Conditioning):** The technology of indoor and vehicular environmental comfort.

50. **Hydronic heating system:** A heating system that uses hot water circulated through pipes to heat a space.

51. **Infrared radiant heating system:** A heating system that uses infrared radiation to heat objects and surfaces directly.

52. **Insulation:** A material that reduces heat transfer.

53. **IoT (Internet of Things):** A network of physical devices, vehicles, home appliances, and other items embedded with electronics, software, sensors, actuators, and connectivity which allows these things to connect, collect and exchange data.

54. **Load calculation:** The process of determining the heating and cooling requirements of a building to properly size HVAC equipment.

55. **Lockout/tagout:** A safety procedure used to ensure that dangerous machines are properly shut off and not started up again prior to the completion of maintenance or servicing work.

56. **MERV (Minimum Efficiency Reporting Value):** A rating system for air filters that indicates their ability to capture particles of different sizes.

57. **Multimeter:** A measuring instrument used to measure voltage, current, and resistance in electrical circuits.

58. **Occupancy sensor:** A device that detects the presence of people in a space and can be used to control HVAC systems.

59. **Ozone depletion potential (ODP):** A measure of a substance's ability to destroy stratospheric ozone.

60. **Planned preventive maintenance (PPM):** A proactive maintenance strategy that involves regular inspections and servicing of equipment to prevent breakdowns.

61. **Pressure gauge:** A device that measures the pressure of a fluid or gas.

62. **Proximity sensor:** A sensor that detects the presence of nearby objects without physical contact.

GLOSSARY

63. **Psychrometrics:** The field of engineering concerned with the determination of physical and thermodynamic properties of gas-vapor mixtures.

64. **Radiant heat:** Heat transferred through electromagnetic waves.

65. **Refrigerant:** A fluid used in HVAC systems to absorb and release heat.

66. **Refrigerant leak:** A leak of refrigerant from an HVAC system, which can reduce efficiency and harm the environment.

67. **Refrigeration cycle:** The process by which an HVAC system uses a refrigerant to transfer heat from one location to another.

68. **Relative humidity:** The amount of moisture in the air compared to the maximum amount of moisture the air can hold at a given temperature.

69. **Return air:** Air that is drawn back into an HVAC system from a conditioned space.

70. **RTU (Rooftop Unit):** A packaged HVAC unit typically installed on the roof of a commercial building.

71. **SEER (Seasonal Energy Efficiency Ratio):** A measure of an air conditioner's or heat pump's cooling efficiency over a typical cooling season.

72. **Sensor:** A device that detects and responds to a physical stimulus (as heat, light, sound, pressure, magnetism, or a particular motion) and transmits a resulting impulse (as for measurement or operating a control).

73. **Smart thermostat:** A thermostat that can be controlled remotely and may include features such as learning algorithms and energy usage reports.

74. **Split system:** An HVAC system that has separate indoor and outdoor units.

75. **Supply air:** Air that is delivered to a conditioned space from an HVAC system.

76. **Sustainable HVAC:** HVAC practices that prioritize energy efficiency, reduced environmental impact, and the use of renewable energy sources.

77. **Thermostat:** A device that controls the operation of an HVAC system by turning it on and off as needed to maintain a desired temperature.

78. **Troubleshooting:** The process of identifying and resolving problems with an HVAC system.

79. **UV (Ultraviolet) germicidal light:** A type of light that kills or inactivates microorganisms such as bacteria, viruses, and mold.

80. **Variable Air Volume (VAV) system:** An HVAC system that controls the amount of air supplied to a space based on its heating or cooling needs.

81. **Variable Refrigerant Flow (VRF) system:** An HVAC system that uses refrigerant as the cooling and heating medium and can vary the flow of refrigerant to each indoor unit based on its needs.

82. **Ventilation:** The process of supplying fresh air to and removing stale air from a building.

83. **VOCs (Volatile Organic Compounds):** Organic chemicals that have a high vapor pressure at ordinary room temperature.

84. **Zoning:** The practice of dividing a building into separate areas, or zones, each with its own thermostat and temperature control.

Bonus Chapter: Fun Facts About HVAC

Congratulations on completing "HVAC for Beginners"! I'm immensely grateful for your time and interest in exploring the intricate world of heating, ventilation, and air conditioning systems with me. Your journey through this comprehensive guide has equipped you with valuable knowledge that can benefit you as a homeowner, DIY enthusiast, or aspiring technician.

As a token of my appreciation, I'm excited to share some fascinating HVAC tidbits with you. Let's embark on a journey through time and technology, uncovering the captivating stories behind the systems that keep us comfortable:

Nature's Air Conditioning

Ancient civilizations were no strangers to the concept of staying cool. Egyptians hung wet reeds in windows, utilizing evaporative cooling as the breeze passed through. This simple yet effective method provided relief from the scorching heat. Meanwhile, Romans ingeniously channeled cool aqueduct water through their walls to beat the summer heat, creating a primitive yet effective form of air conditioning.

The Ingenious Persians

The Persians, masters of innovation, developed windcatchers called "badgirs." These structures captured prevailing winds and directed them downwards into buildings, providing natural ventilation and cooling. Some of these architectural marvels still stand today, a testament to human ingenuity. The badgirs not only highlight the Persian's architectural brilliance but also their understanding of environmental sustainability.

The Iceman's Trade

Before the advent of mechanical refrigeration, ice harvesting was a thriving industry. Large blocks of ice were cut from frozen lakes and rivers during winter, stored in insulated ice houses, and delivered to homes and businesses for cooling purposes. It was a chilly business, but essential for preserving food and providing relief from the heat. This industry laid the groundwork for modern refrigeration and demonstrated early human efforts to control temperature.

A Presidential Perk

The White House, the iconic residence of the U.S. president, received its first air conditioning system in 1930 during Herbert Hoover's presidency. This marked a significant milestone, as air conditioning was still a luxury enjoyed by few at the time. The installation not only improved comfort but also highlighted the growing importance of HVAC technology in modern living.

The Wartime Innovation

World War II spurred advancements in air conditioning technology. Military needs for precise temperature and humidity control in manufacturing facilities led to the development of more efficient and reliable systems. These advancements eventually found their way into homes and businesses, revolutionizing comfort. The war efforts demonstrated the critical role of HVAC systems in ensuring optimal conditions for various operations.

The Heat Pump's Rise to Fame

Heat pumps, capable of both heating and cooling, were invented in the 1940s. However, their popularity soared during the energy crisis of the 1970s when people sought energy-efficient alternatives to traditional heating and cooling methods. The versatility and efficiency of heat pumps made them a popular choice, and their technology continues to evolve today.

The Ozone Layer's Silent Guardian

In the 1980s, the discovery of the detrimental effects of chlorofluorocarbons (CFCs) on the ozone layer led to the Montreal Protocol. This international treaty successfully phased out CFCs, showcasing the power of global cooperation in addressing environmental challenges. The HVAC industry played a crucial role in this transition, developing new, eco-friendly refrigerants that continue to protect our planet.

The Birth of Modern Air Conditioning

In 1902, Willis Carrier invented the first modern air conditioning system to solve a humidity problem at a printing plant in Brooklyn, New York. His invention not only controlled temperature but also reduced humidity, revolutionizing industries that required climate control, such as printing, textiles, and eventually, human comfort.

The Role of HVAC in Space Exploration

HVAC systems are crucial in space exploration. NASA's spacecraft and habitats require precise climate control to ensure the safety and comfort of astronauts. The technology developed for space missions has contributed to advancements in HVAC systems on Earth, particularly in terms of energy efficiency and reliability.

Radiant Floor Heating: An Ancient Luxury

Radiant floor heating, often considered a modern luxury, has ancient origins. The Romans used a system called "hypocaust" to heat their baths and homes. Hot air from a furnace would circulate through empty spaces under the floors and inside walls, providing warmth. This ancient technology is the precursor to modern underfloor heating systems.

Movie Magic and Air Conditioning

The 1920s saw a boom in air-conditioned movie theaters, with the famous Roxy Theater in New York being one of the first to install air conditioning. This innovation made movie theaters a popular summer destination, offering a cool escape from the heat and transforming the movie-going experience.

The World's Largest HVAC System

The world's largest HVAC system is located at the Grand Mosque in Mecca, Saudi Arabia. This massive system is capable of cooling the entire mosque, which can accommodate up to two million worshippers during the Hajj pilgrimage. The system ensures a comfortable environment even in the extreme heat of the desert.

The Skyscraper Wind Effect

Tall buildings create a unique wind phenomenon called the "skyscraper wind effect." This can lead to increased wind speeds at street level, making it crucial for HVAC designers to account for wind loads on outdoor units.

The Igloo's Ingenious Design

Igloos, built by the Inuit people, are surprisingly effective at insulation. The unique shape and snow construction create a warm and comfortable shelter in freezing temperatures.

In Conclusion

I hope these fun facts have sparked your curiosity and deepened your appreciation for the intricate systems that keep us cool in the summer, warm in the winter, and breathing clean air year-round. Thank you once again for reading "HVAC for Beginners." Your journey doesn't end here—continue exploring, learning, and enjoying the world of HVAC.

Bonus Quiz

Ready to prove your HVAC expertise?
Scan the QR code below to take our quiz and earn a certificate of completion. Good luck!

Click here for ebook readers:

https://forms.gle/jm3gwNVaN7LPx1uDA

Made in the USA
Middletown, DE
10 September 2024